A New Way to Pay

A New Way to Pay

Creating Competitive Advantage Through the EMV Smart Card Standard

ANEACE HADDAD

GOWER

Published by
Gower Publishing Limited
Gower House
Croft Road
Aldershot
Hants GU11 3HR
England

Gower Publishing Company
Suite 420
101 Cherry Street
Burlington,
VT 05401-4405
USA

British Library Cataloguing in Publication Data
Haddad, Aneace
A new way to pay: creating competitive advantage through
the EMV smart card standard. – 2nd ed.
1. Smart cards 2. Customer services 3. Marketing
I. Title II. Haddad, Aneace. Using smart cards to gain market
share
658.8'83

ISBN 0 566 08688 3

Library of Congress Cataloging-in-Publication Data
Haddad, Aneace.
A new way to pay: creating competetive advantage through the EMV smart card standard / by Aneace Haddad.
p. cm.
Rev. ed. of: Using smart cards to gain market share. 2000.
Includes index.
ISBN: 0-566-08688-3
1. Customer services. 2. Smart cards. 3. Marketing. I. Haddad, Aneace. Using smart cards to gain market share. II. Title
HF5415.5.H24 2005
658.8'83--dc22

2005005499

Typeset by IML Typographers, Birkenhead, Merseyside.
Printed and bound in Great Britain by TJ International Ltd, Padstow, Cornwall.

Contents

List of Figures and Tables

FIGURES

TABLES

Preface

The world's payment infrastructure is going through a major upgrade to EMV, the smart card standard mandated by Europay, MasterCard and Visa to combat fraud. In addition to fraud, EMV is also being used to modernize payment card products, as a way for banks to deal with rapidly increasing losses due to shrinking fees and commissions, losses which are often far greater than those due to fraud. UK banks are losing £400 million per year due to card fraud, but they are losing £1 billion per year due to customers switching card companies, seeing no real difference between cards other than the rates. Also, UK authorities will soon be forcing banks to reduce the transaction fees charged to merchants, judging that the price of the service is much higher than its cost, resulting in a loss of revenues estimated at another £1 billion. Commoditization may soon be costing UK banks £2 billion per year, a problem at least five times as great as fraud.

This book is about using EMV to break out of the commoditization box. It is about transforming the basic payment function into a richer, more memorable payment experience – getting cardholders and merchants to see payment as something new, exciting and different, so that they will focus on the added value you provide rather than the cost you represent; getting merchants excited about EMV so that they in turn create excitement around your payment brand in their stores.

Is an EMV deployment successful when all of the technical problems have been eliminated? When fraud is under control? When the budget is respected and the project completed on time? In this book, the definition of success is far more ambitious. The move to EMV is such an important event that we as an industry need to be far more demanding. If a deployment is successful according to all of the prior criteria, but fails at creating differentiation, fails at positioning payment as an integral and strategic tool that justifies merchants continuing to pay ongoing transaction fees and commissions, fails at getting merchants to look at payment as anything but an intolerable expense that needs to be cut, then the migration to EMV cannot be considered truly successful.

Aneace Haddad
2005

CHAPTER 1

The Threats – Commoditization and Fraud

How can I get more benefit from EMV than simply becoming compliant, just like all the other banks?

Managing Director, European bank

Banks throughout the world are upgrading their payment infrastructure to become compliant with EMV, the smart card standard driven by Europay, MasterCard and Visa. Initially, EMV was developed primarily to combat losses due to rising levels of card fraud, such as counterfeit and lost or stolen cards. In the UK alone, payment card fraud has recently been running at over £400 million annually, approximately half of which could be eliminated with an EMV payment system. Today, in addition to combating fraud, EMV is also being used by banks to modernize and improve their payment products and services. This is critical for an industry facing rapidly increasing losses due to margin erosion and shrinking fees and commissions, losses which are often far greater than those due to fraud.

Commoditization is a term which describes what happens when a company's products and services are so similar to the competitors' that the only real difference is the price. The effects of commoditization are being felt by banks everywhere. Because the market is focused on price, consumer and merchant advocacy groups all over the world are successfully arguing that card fees and commissions are too high and are increasing, while the actual cost of providing the services is going down. Banks are being forced to cut their revenues. For example, UK banks face a forced slashing of interchange fees by up to £1 billion a year, if the Office of Fair Trading gets its way.[1] That's more than twice the rate of fraud. If you add to that recent reports that another £1 billion of revenue is being lost every year due to cherry-picking customers who move perpetually from issuer to issuer, shopping around for the lowest-cost credit card and enjoying enticing introductory offers,[2] you will find that UK banks risk losing over £2 billion annually due to commoditization, resulting in a problem at least five times as great as fraud.

The business case for EMV is built around combating fraud. It works when fraud reaches the levels experienced in France in the 1980s and more recently in the UK and Malaysia. But the payback is long. Given the current rate of fraud, it could take UK bankers up to three years to recover the cost of EMV. The business case becomes far more powerful when EMV is also used to combat commoditization. Using new technology to get a £400 million fraud problem under control is much easier to justify if the same technology is used instead to combat fraud plus commoditization, potentially a £2.4 billion problem. Even a relatively small impact on commoditization could help get a positive payback on EMV in the first year.

When you are in front of a merchant to pitch your co-branded services, or your payment-processing services if you are an acquirer, how often have you found that one of the major themes driving the negotiation is that your card fees are too high? That your competitor provides essentially the same service for a lower cost? That differentiating your products from your competitor's has become a very complicated process and requires more and more explaining? That, in the merchant's mind, you represent an expense which must be driven down at all costs, and that any savings will immediately go to the merchant's bottom line? Worse, that the merchant increasingly compares the cost per store of processing card transactions to services like utilities and rent? In other words, that you are increasingly boxed in as a provider of commodity services?

This book is about breaking out of the commoditization box. It is about getting your customers, whether they are cardholders or merchants, to see payment as something new, exciting and different, so that you can get them to focus on the added value you provide rather than the cost you represent. It is about using EMV to transform your payment infrastructure into a tool that helps merchants solve important problems that are much bigger than the costs of your services. It is about getting merchants excited about EMV so that they in turn create excitement around your payment brand in their store. Your cardholders will quickly see this. They will be encouraged to use your card more often, and if they don't have one yet, they will want one.

MIGRATION TO EMV: GETTING A FASTER RETURN ON INVESTMENT

Starting from January 2005 in Europe, and January 2006 in Asia, banks and merchants will have to bear all losses incurred in fraud cases involving non-EMV cards and non-EMV terminals, which could have been avoided thanks to EMV. The liability for fraud will essentially shift from the card associations, Visa and MasterCard, to the party that is non-compliant, either the bank or the merchant, or both. Non-compliance could be costly. To prepare for the upcoming shift of fraud liability, banks worldwide have already begun investing in EMV migration.

Magnetic strip cards are easy to duplicate, and well known techniques exist for thieves to skim credit card details and PIN numbers to create fake cards. A simple magnetic strip card reader can be used by a clerk to swipe a customer's card while the customer is not looking. More sophisticated techniques include pinhole cameras installed above ATMs that film the customer's PIN code and card details, which are then used to create duplicate cards. Similar techniques are spreading across the world.

The problem with skimming is that it is so easy to do and so hard to stop. Credit card skimming is now a major financial crime across the world and exceeds the combined total for cases of lost or stolen cards. Since most cloned credit cards are used outside of the country in which they were issued, tracking and resolution of the stolen details is very difficult. Card skimming is attractive to criminals, as the cards can be used for longer before the cardholder realizes and cancels their card, while stolen cards are often cancelled immediately. Petrol stations and restaurants are favourite places for card skimmers to operate. Restaurant waiters have been caught with card-skimming devices hidden in their aprons, while the ATM lobbies of banks are sometimes rigged with illicit card-reading devices that store the details of cards used to gain access.

EMV was designed to eliminate these problems. As fraud grows, and the liability shift approaches, EMV migration becomes unavoidable.

Banks in France converted their payment infrastructure to chip cards in the late 1980s, when fraud rates reached close to 0.30% of the total cost of goods purchased. In less than five years, fraud was down to an acceptable 0.08%, the current level of fraud in the US, which banks are comfortable with. In less than ten years, French bank card fraud was down to an almost insignificant 0.02%, the lowest rate anywhere.

UK banks expect to complete their migration to chip and PIN by 2005 in order to achieve similar results. Prior to adopting chip and PIN, UK fraud rates had been approaching 0.30% of transaction volume, considered to be a crisis threshold requiring critical action; this represented a total loss of £400 million, approximately half of which could be eliminated with an EMV payment system. UK banks expect to invest over £1 billion in upgrading their payment systems to EMV standards, an average of £26 per cardholder. Given the current rate of fraud, it will take several years to recover the cost of migration. EMV migration is not only costly, it's also difficult.

Malaysia was the first country in the Asia-Pacific region to mandate a full migration to EMV by all of the country's banks in response to high fraud levels. Total card fraud, including counterfeit and lost or stolen cards, has also been estimated at up to 0.30% of transaction volume, the same critical threshold reached in France in the 1980s and in the UK more recently. Just six months prior to the government-mandated liability shift deadline, MasterCard's Jim Cheah, senior country manager for Malaysia, said that eight of Malaysia's 19 MasterCard issuers and 60% of 50 000 participating merchants had yet to migrate to EMV. Cheah said he was optimistic that the majority of these card issuers and merchants would be compliant by the deadline. The terminal conversion for the merchants would be effected through stages because of the numbers and training involved. To encourage banks and merchants to move faster, Cheah reminded them that the adoption of smart card technology would drastically reduce credit card fraud cases, pointing to France, where there were 80% fewer fraud cases after three years of chip card implementation.[3]

As more Asian countries adopt the EMV standards, the fraudulent elements there are expected to shift to regions such as Australia where the tough new standards are yet to be introduced, and where banks have yet to establish a clear timeline for EMV compliance.[4] Even in the US, where credit card fraud is relatively under control, the decision by US credit card issuers not to adopt EMV smart card standards is expected to make US cards a target for fraud.[5]

So why aren't all bankers migrating to EMV at the same time?

Many bankers around the world have been watching closely as their UK colleagues upgrade their payment infrastructure to chip and PIN. They have seen that the investment is high and that the primary rationale for EMV, that it will slash fraud losses, is not quite enough to justify the effort.

Given the cost of EMV and the time it can take to recover one's investment, it might be better for many banks to take a wait-and-see attitude. If fraud remains under control, perhaps it is best for a bank to swallow the liability. But if fraud does spiral out of control quickly, the bank would need to be confident of its ability to deploy an EMV infrastructure even more quickly. If fraud explodes, how much exposure can a bank withstand during the minimum 12 to 24 months necessary to replace all cards and terminals? This is not an easy decision to make.

The hard financial cost of fraud is only one way of looking at fraud risks. There is another way, which is more difficult to quantify and which is actually scarier. Citibank in the US has been running an advertisement showing an overweight grandfatherly man getting a call from his bank asking if he just bought two surfboards in Maui. The man looks more like a factory worker on a limited budget than a surfer who escapes to Maui whenever he gets the opportunity. Another ad shows a conservative-looking middle-aged housewife sitting in a tattoo parlour, looking very much out of place, getting a big tattoo on her arm. The ad says, 'This looked suspicious to us too.' Fraud and identity theft are beginning to be felt by the general public. Financial institutions must show that they are dealing with the problem and that customers need not worry. The real risk of fraud is the public's loss of confidence in plastic payment methods. A single high-profile fraud occurrence, which in itself represents a tiny hard financial value in actual quantifiable losses to a single bank, could be enough to trigger a much larger impact on payment systems through loss of confidence.

Here's an example. In response to high rates of credit card fraud in Malaysia and neighbouring countries, Korea's Samsung Card began advising clients not to use its credit cards in south-east Asia. In early 2003, Samsung Card reported that fraudulent card use in the region comprised 59.7% of its total overseas fraud. Malaysia led the list at 22.8%, with the Philippines second, at 15.3%.[6] I am sure that Malaysia is a favourite vacation spot for Korean tourists, but certainly not at that rate.

When fraud grows quickly, there is a real sense of loss of confidence in plastic payment methods. The smooth flow of payments can be seen as a key factor in increasing consumption and growing the economy, so loss of confidence in plastic could have serious consequences. Hence the government mandate in Malaysia to eliminate fraud by moving to chip. Since chip migration is costly, and the business case based on fraud alone is not quite enough to produce an acceptable return on their EMV investment, Malaysian financial institutions and merchants are looking at enhancing the payment experience with new features in order to fully benefit from the smart card's greater capabilities. The idea is to use EMV to combat commoditization in addition to fraud, rather than use EMV to combat fraud alone. According to MasterCard's Jim Cheah, 'Merchants could use the EMV-compliant terminals for data mining purposes or to enhance customer service through discounts, cash back or free gifts'.

Bankers in many parts of the world are seeking ways to improve the business case for EMV in order to obtain a more acceptable return on investment (ROI) and decision-makers are asking hard questions before approving major EMV budgets.

GETTING MORE BENEFIT FROM EMV

Why should I spend a fortune just to have the same capabilities as everyone else? Does it really make sense to my shareholders that I will be spending so much money just to be able to step up to a level playing field? How can I do better with my investment? How can I invest smarter than everyone else and get more for my money than my competitors? What can I do to leapfrog my competitors and get more value out of EMV migration than everyone else?

The answers to these questions can be found by looking at EMV migration within a larger context. What other core business problems can EMV solve for bankers, in addition to reducing fraud? Commoditization is high on the list of pain factors keeping bankers awake at night.

COMPETING BETTER WITHOUT SIMPLY SPENDING MORE AND CHARGING LESS

According to Singapore's *Business Times*, Asian bankers lament the irrational pricing, the jostling for consumers' attention, and the cherry-picking customer who flits from bank to bank. Margins are on the rise but face a ceiling with intensifying competition in what is already the lowest-margin environment in the region. Competition has been dominated by pricing, discounts and promotions. Credit cards – one of the hotly contested products – now mostly come free along with lucky draws for a Jaguar (from OCBC Bank) or Ferrari (from DBS Bank) as well as discounts at merchants. And banks have to spend more to grab attention. HSBC's aggressive advertising, which leads the bank to spend tens of millions yearly, is due to 'the need to get noticed'. Consumer banking is huge – and customers tend to cherry-pick. These are the customers who choose the lowest-priced product from each bank and expect royal treatment thrown in.[7]

Low-interest credit card offers are easy to find all over the web and in your letterbox. Here are just a few offered recently.

> 'Citi Platinum Select – 0% intro APR first 9 months, 9.49% thereafter. No annual fee.'
>
> 'American Express Blue – 0% intro APR first 15 months, 7.99% thereafter. No annual fee.'
>
> 'Chase CashBuilder – Low 9.24% APR, 0% on transfers for 12 months. Earn 2% cash back. No annual fee.'

In the US, where fierce rate wars have resulted in the widespread availability of reduced credit card rates, no annual fees, and special low introductory and balance transfer rates, economists have indicated that increased competition is costing the credit card industry about US$30 billion per year.[8] In 1990, only 6% of card balances were at interest rates below 6.5%, and 93% were above 16.5%. By 2002 almost three-quarters of all outstanding balances were at interest rates below 18%, while 15% of balances were at interest rates under 5.5%. At the other end of the spectrum, only 24% of outstanding balances had interest rates above 18%.

Credit card interest rate margins have hit their lowest levels in years. In many countries, card issuers inundate the market with introductory offers that promise no interest for a year or more. Customers have learned how to switch between credit card issuers so that they can perpetually benefit from low-cost introductory offers.

In the UK, customers who continually switch between 0% finance deals on credit cards to keep their debts interest-free are costing credit card firms around £1 billion a year in lost revenue and reduced margins. The introduction of chip and PIN could result in a consolidation of cards used by consumers, due to the need to remember, or change, multiple PIN codes. The number of cards issued is expected to continue to increase, although the number of cards actually used by consumers will decline. It is likely that a number of affinity cards, which are not primary or even secondary cards in the wallet, will be cut up or not used as a result of chip and PIN. In addition, the expected reduction in interchange fees will put pressure on revenues from these schemes.

Not only are interest rates the lowest ever, but banks now have to spend more on advertising and marketing just to get noticed. In the US, individual card issuers and the major card networks (Visa, MasterCard, American Express and Discover) spend close to US$11 billion each year on card advertising. That comes out at around US$57 of annual advertising spending for each credit card-carrying American. Any impact on advertising effectiveness can quickly yield substantial savings.

Even in Thailand, where the credit card market is growing briskly and many potential customers still don't have cards, issuers must resort to doling out gifts and prizes to lure customers their way. During Money Expo 2004, Kasikornbank offered a small refrigerator to its first 4000 applicants.[9] When discussing Kasikornbank's free refrigerator offer with a credit card executive at a leading Hong Kong bank, the executive laughed: 'That's nothing, we sometimes have to give new cardholders a free colour television!'

Card executives across many different countries and regions, in Europe, Asia, the Middle East, Latin America and the US, regardless of GDP, confirm over and over again that the cost of acquiring a new cardholder in their particular market is easily in line with the often-cited US figure of US$120 per cardholder. This is true in mature markets like the US, the UK and Hong Kong – where the market for new credit card customers is virtually nonexistent and competition consists primarily of stealing customers away from other banks – as well as growth markets where competition for new credit card customers is intense, such as Turkey and Thailand where payment cards are a new and rapidly growing phenomena, or France and Germany where there are still very few credit cards and the market is poised to take off. Saving just a few dollars on each new cardholder can quickly add up to big numbers.

RAISING NON-INTEREST INCOME WHEN THE MARKET IS DRIVING FEES DOWN

This is another important commoditization issue. With interest rates at record lows, how can my bank fully maximize its non-interest income? Analysts regularly attribute high bank profitability in recent years to the significant growth of non-interest income – the revenue that banks earn from areas outside their lending activities. Any income that banks earn from activities other than their core lending business or from their investments is classified as non-interest income. This type of income is often referred to as 'fee income' since fees constitute the majority of non-interest income. Over the last 20 years non-interest income has transformed from a supportive role into a major contributor of bank revenue. Most importantly, banks that increase non-interest income could reduce risk through greater diversification. Non-interest income is typically described as more steady or stable than interest income.

Banks want and need non-interest income, but the market is pressuring banks across the world to lower their fees. In response to market pressures, many card issuers have to waive the annual fee for the first year and sometimes beyond. Some even have to offer top customers a lifetime of waived annual fees.

Here is some practical advice provided by a credit card customer for all to see on his website:

> When I call to ask for an annual fee to be waived, my experience has been that 95% of the time they do it! There have been a few times when the bank hasn't waived the fee, so I've simply closed the account. I have too many credit card options to pay any annual fees. That's a position you want to be in. You don't want to be at the mercy of one bank. No matter your situation, bad credit, good credit, or other, you should always call to see if you can make the bank waive your annual fees. Be sure to keep an eye on every transaction in each month's credit card billing statement. If you see an annual fee charge, call the bank and get that fee waived!

Commoditization indeed.

Merchants have also developed a sophisticated understanding of card fees and their impact on profitability. They are also using the web for tips on how to deal with fees. According to the website of the National Association of Convenience Stores, credit and debit card fees cost the average convenience store US$24 265 in 2003, a figure approaching the average per-store pretax profit of US$30 700. With razor-thin margins for convenience stores selling motor fuels, the situation is even worse, as retailer associations are finding that credit card companies often make more profit on a gallon of gasoline than the retailer selling the gasoline. The National Association of Convenience Stores is working with First Data Corporation to convince card processors to charge a fixed cost per transaction rather than charging a percentage of the sale. As the dollar value of the transaction grows (such as with the rising price of gasoline), the card processing fees remains the same.[10]

Visa and MasterCard put a lot of effort into explaining and justifying their fee structures to merchants and government authorities in order to protect their existing fee revenue streams. These efforts increasingly fail. Worse, merchant groups and government authorities in some regions have been encouraged by changes imposed on banks in other regions. The current trend is not favourable to banks.

In 2002, Visa was forced by the European Commission to lower its cross-border interchange fees for credit and offline debit cards. By 2007, it must match the French level of 0.7%, chosen by the European Commission as a benchmark, everywhere in the euro zone.

In the US, Visa and MasterCard recently experienced what has probably been the biggest forced reduction in fees in the history of transaction processing. In April 2003, after nearly seven years of legal battles, MasterCard and Visa finally put an end to an anti-trust lawsuit brought by more than five million retailers, including Wal-Mart, Sears, Safeway and Circuit City. The two card companies agreed to pay US$3 billion and to lower the interchange fees on debit cards. As a result, analysts estimate that banks will see margins on merchant-acquiring activities contract by about 1.5% to 2.0%, which is expected to save merchants a total of between US$63 billion and US$100 billion through 2010.[11]

Also in 2003, but on the other side of the globe, Australia's Reserve Bank mandated a major reform of credit card fees and ordered banks to slash interchange fees by about one third. From an average of around 0.8% before the reforms, Australian banks now charge approximately 0.5% on Visa and MasterCard transactions. The Reserve Bank argued that the banks had been charging these fees above the real cost of the service and using the excess to subsidize their loyalty schemes. One year later, in July 2004, the Reserve Bank's initiative had succeeded in cutting bank revenues by an annual rate of AU$430 million.[12]

In the past, Australian banks often financed their customer rewards programmes using the interchange fee. Cutting fees means that there is no longer any room for banks to offer a traditional catalogue-based rewards programme, given the high cost of managing a catalogue, generating rewards statements, keeping a rewards inventory, processing and delivering rewards requests to customers and managing a call centre. The overhead involved in operating a rewards programme often represents close to half the total programme budget, with the rest of the budget used to pay for the actual rewards, gifts, air miles and so on.

The impact of the Reserve Bank's reforms was immediate. In a letter to its cardholders, Westpac has advised of a 'repricing' by which the value of points earned has been devalued by 12%. While Westpac has maintained its conversion ratio at one point for every dollar spent, the points will not buy as much. 'From October 1 you'll need 14 000 Altitude points to redeem a $100 David Jones voucher, instead of the current 12 500 Altitude points needed,' the letter says.

Of course, banks have found ways to sidestep the Reserve Bank's crackdown on interchange fees. National Australia Bank hooked up with American Express to issue a new card that was not covered by the Reserve Bank's mandate and now claims it is the best credit card rewards programme it has ever offered.[13] Its launch follows similar moves by ANZ and Westpac, which have introduced cards with Diners Club and American Express offering comparatively generous reward schemes. In the case of Westpac, the bank's letter makes no bones about the fact that its repricing is aimed at migrating points-hungry customers to its Altitude American Express 'companion' card.

The Reserve Bank's payments board is investigating whether these arrangements break the spirit and intent of the interchange reforms. Visa and MasterCard are sure to fight back as well. According to some experts, the Reserve Bank's reforms so far have resulted in higher annual card fees and lower rewards, especially with frequent flyer points.

Something quite similar is now happening in the UK. In November 2004, Britain's competition watchdog, the Office of Fair Trading, found that banks had been overcharging retailers for processing credit card payments and that the high fees were ultimately borne by consumers. Analysts at Morgan Stanley believe that the OFT could be prepared to insist that fees be cut by more than two thirds, from the current 1.1% of the total cost of goods purchased to 0.35%. That would cost the credit card industry £1 billion a year in lost revenue. Some analysts believe that if interchange fees are cut by the Office of Fair Trading, other incentives such as Air Miles and Nectar points will also disappear.[14] That is precisely what happened in Australia.

A slightly different angle on the story is happening in France. In July 2004, the European Commission accused Groupement des Cartes Bancaires (CB), the operator of the French bank card system, and nine of its largest member banks of conspiring to restrict new entrants and stifle innovation in the French cards market by imposing complex fees and regulations that prejudice rival issuers. The end result will probably be lower fees.

This is not only true in mature payment card markets, but also in new growth markets for credit and debit cards. Over 600 million bank cards had been issued in mainland China by September 2003, up 24% in one year, and this growth is bringing changes in merchant fees for card acceptance. From 1 March 2004, card issuers were permitted to levy a settlement fee of zero to 1.6% of the value of an electronic payment at the point of sale, instead of the previous 1 to 2% of total payment. In a move to catalyse the use of bank cards and e-payments, the highest rate will be paid by hotels, restaurants and entertainment venues, while state hospitals and schools will have free access to electronic payments.[15]

In addition to Visa and MasterCard efforts to convince the market of the value of payment and the reasoning behind the current fee structures, significant effort is also needed to enhance the traditionally limited payment experience, with new features offering new value to merchants and cardholders. That is the best way to break the perception that payment is a commodity.

Since payment is increasingly perceived to be a commodity, and the market is putting increasing pressure on banks to reduce settlement fees and other payment processing costs, what can be done to reverse that trend? What can be done to make payment more valuable, more exciting, with new services which are not the target of government watchdogs intent on bringing the price of payment in line with its perceived status as a low-cost commodity? Lost fee revenue of £1 billion a year in the UK is more than twice the £400 million volume of fraud losses and more than four times the volume of losses directly avoidable through EMV chip technology. In Australia a single event, the Reserve Bank's mandated fee reductions, has reduced annual bank revenues by AU$430 million, while annual fraud in Australia is estimated at around AU$100 million, approximately half of which is attributable to skimming and therefore easily avoidable with EMV. In the US, another single event, the anti-trust lawsuit against Visa and MasterCard, could represent lost fee revenue of over US$10 billion per year, while annual credit card fraud in the US is estimated at around US$3 billion. An estimated 60% of those losses are due to thieves skimming information from the card's magnetic strip. Erosion of transaction fees in all parts of the world is always a much bigger problem than fraud.

Reduced fees are only one part of the commoditization problem. In addition to fee reforms driven by industry watchdogs, resulting in substantial reductions in fees charged to merchants, commoditization has also resulted in rate wars, leading to the slashing of interest rates and annual fees charged to cardholders. When examined overall, losses due to commoditization are at least five times greater than total losses due to payment card fraud, sometimes over ten times greater. Commoditization simply dwarfs fraud (see Figure 1.1). As banks discover that they can use EMV to address both fraud and commoditization, as opposed to fraud alone, for a virtually insignificant increase in the total cost of EMV migration, it's not surprising that more and more banks are choosing to do precisely that.

Wherever EMV is growing, banks are looking at ways to get more value out of EMV by using the technology to combat commoditization in addition to fraud. EMV is being used to offer new features and services which can help generate new fees and increase non-interest income, in many cases replacing fee reductions mandated by

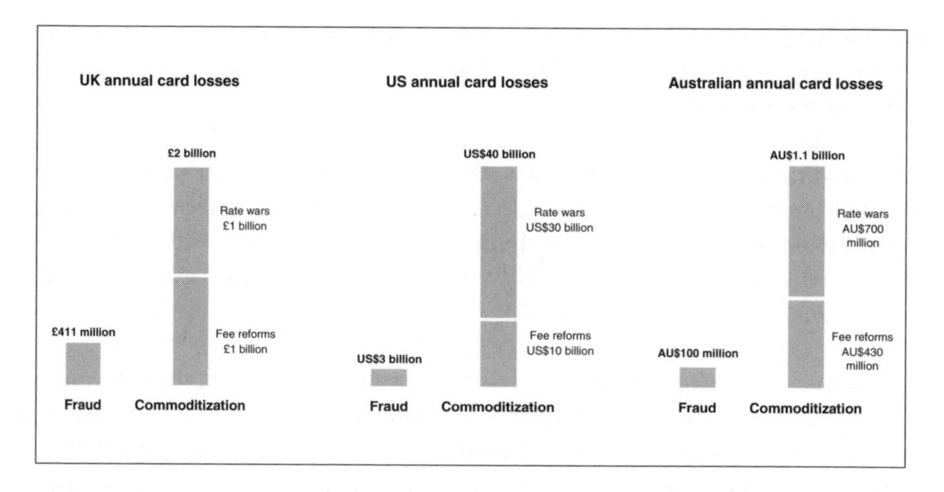

Figure 1.1 Losses due to commoditization versus losses due to fraud

government bodies or forced by competition. In countries where fraud is high and the need for EMV is growing, EMV chip cards are also being used to eliminate most of the operational overhead involved in traditional rewards programmes. Taking the case of Australia, where banks have become accustomed to allocating interchange fees of around 0.8% of the total value of customer purchases to a rewards programme, it is likely that many programmes burn up to half of that in operational overhead, leaving a little more than half for the actual rewards provided to customers. Through the use of EMV, banks can eliminate much of the unnecessary burn and allocate 0.5% to the rewards programme, with little or no impact on the value of the rewards given to customers. Also, by using EMV's full payment features, including customer-centric enhancements, banks can invite merchants to offer surprise promotions to their most frequent customers, targeting their paper incentives more efficiently using detailed behaviour data stored in the chip. In this way, generous discounts and rewards are offered to customers and are paid by merchants, not out of the interchange fees.

CREATING DIFFERENTIATION IN AN INCREASINGLY COMPETITIVE MERCHANT-ACQUIRING AND PAYMENTS-PROCESSING MARKET

Merchant-acquiring has long been regarded as the least attractive part of the payments business, generating lower margins than card issuance, and seen as strategically less important. Banks have traditionally focused on their payment brand, card issuance and transaction growth, paying little attention to merchant acquisition. Many banks have withdrawn completely from the acquiring value chain, especially in the US, and for others it is a low-priority business attracting less than 5% of the total global investment in card businesses.[16]

That was yesterday. The market is changing.

Major US acquirers are aggressively expanding outside the US, large banks are once again expanding their own acquiring activities, and smaller European processors have merged to create a larger entity capable of competing more effectively. All are looking for ways to differentiate their products and services and be better competitors on the basis of features and benefits rather than price.

First Data International aims to grow the non-US segment of its revenues from the current 25% to 50% in five to ten years, on the basis that cards will be the dominant payment medium. First Data International's president, Pam Patsley, sees smart cards, loyalty and wireless innovations as the most likely 'emergent payment technologies' to take off in the next five years. With 'the basic elements of card transaction processing ... becoming very much a commodity', Patsley says, 'competition is driving the need for differentiation, which makes card issuing a "scale" business.' Processor consolidation within Europe is another major trend for FDI, which believes that:

> In the next five years, there will be a dramatic reduction in the number of card processors in Europe. On a truly global basis, probably no more than ten card issuing [firms] and perhaps three or four global processors will exist by 2007. Consolidation will also be driven by the increasing interest in pan-European activity, and the challenge for new entrants [to the processing market] will be one of achieving scale quickly enough to be competitive.[17]

In just the first six months of 2004:

- First Data, already the largest independent payments processor in the world, acquired Delta Singular, a provider of payment-processing services in Greece, the Middle East and the Balkans.
- Global Payments acquired Prague-based MUZO, the largest independent payment processor in the Czech Republic, holding approximately 50% market share. According to Global chairman Paul Garcia, the move will allow the company 'to expand our presence to the European payments market'.
- Nova Information Systems signed three deals to purchase European processors. According to Nova, the acquisitions underscore a shift in European banks' attitudes toward outsourcing and its chief executive says that the company plans to pounce on that development.[18]

- Royal Bank of Scotland purchased US merchant-acquirer Lynk Systems. Following the acquisition, RBS became the third largest acquirer in the world. Another RBS acquisition in the US, Charter One, already had agreements with retailers including Wal-Mart, Starbucks and Kroger to offer in-store banking, and Royal Bank expects to begin issuing co-branded credit cards to Kroger customers.
- Bank of America acquired National Processing Inc. and plans to combine it with Bank of America Merchant Services to create the second largest bankcard merchant-acquirer in the US, with nearly US$250 billion in annual processing volume. The transaction is expected to allow Bank of America to compete more effectively in the electronic payments business by creating immediate scale.

At the same time that competition in acquiring and processing is growing rapidly, EMV migration is expected to boost outsourcing business for payment processors as banks increasingly choose not to take on full-scale EMV integration efforts in-house. The potential of lucrative long-term EMV processing contracts is driving even more competition amongst processors and outsource providers.

BANKS USING EMV TO REDUCE FRAUD AND COMBAT COMMODITIZATION

Strategic-minded banks throughout the world are turning the liability-shift mandate into an opportunity to create new payment products which significantly increase revenues and market share, boost non-interest income and help reduce the cost of acquiring new cardholders and merchants. In the same way, strategic-minded payment processors and merchant-acquirers are looking at creating new payment services to differentiate them from their competitors, fight against commoditization and win more outsourcing business from banks and retailers.

The liability shift is an excellent opportunity to build a full-featured EMV infrastructure from day one. This is the fastest and least expensive way to break the perception that payment is merely a commodity service. Today all EMV cards allow additional data to be processed within the chip, not just fraud prevention data. Even the lowest-cost EMV cards and terminals have enough memory and processing power to enhance the payment experience by including features such as instant delivery of surprise offers, welcome gifts and targeted messages, immediate calculation of points or cash back, and easy redemption of rewards at the point of sale. Banks are finding that adding these features to the payment services they already provide has little if any impact on the card and terminal equipment costs or even personnel costs related to providing these services. In one example, a bank has only had to assign two people on a part-time basis to operate enhanced payment services for over 80 000 merchants.

Banks have found that EMV can provide valuable differentiation in an increasingly competitive market, with little impact on the total cost of chip migration. An EMV infrastructure which includes customer-centric enhancements, in addition to basic fraud features, costs little more than basic EMV, but offers far more benefits for cardholders, merchants and bankers.

CUSTOMERS ARE ATTRACTED TO PAYMENT CARD FEATURES THAT SIMPLIFY LIFE

Processing rewards directly at the point of sale is a powerful enhancement to the payment experience and an effective way of making the payment experience more memorable. Customers get their points balance directly on their credit card receipt, instead of waiting to receive their statements by mail, and they can redeem their rewards at the point of sale to pay for purchases. Points are just another way of paying. This eliminates the high cost of maintaining a catalogue of reward items, keeping the rewards stocked in a warehouse, processing requests for rewards and sending the rewards by mail to customers. Customers can enjoy their rewards whenever they like, instantly. With a full-featured EMV infrastructure, customers can choose to pay by credit, debit, points or coupons. It's all money.

Customer-centric features make an EMV card more attractive to cardholders, who can get rid of the loyalty cards and coupons stacking up in the kitchen or making their wallets fat, and can stop worrying about missing out on rewards and valuable discounts. This helps banks recruit new cardholders at lower cost – an important factor in a competitive market where the average cost of recruiting a cardholder is estimated at over US$120. One issuer succeeded in recruiting a million new cardholders in eight months, almost twice as many as anticipated, without substantially increasing its advertising and marketing budgets. In this example, for a credit card portfolio of one million new cardholders, the estimated savings on cardholder recruitment costs could be valued at between US$30 million and US$60 million.

By enhancing the EMV payment experience with customer-recognition features, a retailer is able to recognize a new customer who has not yet used their card at that retailer's outlet, or a customer who has not been to the store for some time. The credit card terminal can prompt the sales clerk to offer a welcome gift to those customers. When the display flashes 'give welcome gift', the clerk reaches under the counter and offers the surprise gift to his customer. Surprise gifts or special reserved promotions can be triggered instantly at the moment of purchase to delight frequent customers, giving the highest-value gifts to the best customers and making the payment experience more memorable. These are simple payment enhancement features which help banks use EMV to its fullest and help get retailers involved in the migration to chip. The credit card receipt can also include a short list of 'My Reminders', to help customers remind themselves of special events, such as a spouse's birthday. Some merchants would appreciate the ability to offer special promotions to those customers.

When one looks at the amount of data available today at the moment of purchase, within the card and the terminal, and asks the question, 'What can I do to enhance my customer's payment experience with all this data?', the ideas appear limitless. This line of reasoning is far more powerful than the more traditional, 'How do I convert my magnetic strip loyalty programme to chip?', which is a far more limited thought process that will most likely cause one to miss out on many simple and powerful new payment features.

Of all the new payment features available for chip cards, points, coupons and receipt-reward messages are closest to the pure payment function. These features build upon, enhance and improve the customer experience at the moment of payment. The customer's card is inserted in the payment terminal, the amount of the transaction is keyed in, and the credit card receipt is printed, including the traditional credit or debit card information as well as the customer's rewards balance, special promotional messages and any other enhancements offered by the card issuer and the merchant.

ENHANCED EMV HELPS BANKS GET A POSITIVE ROI OUT OF THEIR EMV INVESTMENT

Payment enhancement features are generally available at a relatively small incremental cost on top of the cost to migrate to EMV. Because payment enhancement data is already supported within most EMV cards, banks can choose to use either low-cost 'single application' EMV cards or the more sophisticated 'multi-application' cards with no loss in functionality.

So called 'single application' EMV cards are the lowest-cost payment cards on the market. They are produced by all the major card manufacturers and include well known and popular cards such as Gemplus's MPCOS, Oberthur's VSDC, or G&D's StarDC EMV card. All of these cards support standard ISO commands to securely read and write data in encrypted files that are separate from the fraud-related data. No modification is needed to the operating system and no new applets need to be added. At the same time that the cards are personalized with the customer's name and account details, to prepare them to be mailed out to customers, a data file needs to be allocated during the same personalization process, which the terminal application will later use to store payment enhancement data, counters, coupons and so on.

With multi-application Java and Multos cards, the data management applet that stores payment-enhancement data in the chip is generally already loaded to the card when it is manufactured in Read Only Memory (ROM), or when it is personalized in Electronically Erasable Programmable Read Only Memory (EEPROM). Both of these methods eliminate the need for financial institutions to invest in multi-application card management systems that manage the dynamic addition of new applets to cards, which is one of the costliest features of multi-application chip cards. The most successful smart payment card deployments have avoided things like 'post-issuance download' – which translates to the ability for customers to load new applets to their cards via card readers attached to their home PCs, or via special kiosks or ATMs – and have concentrated on providing enhanced payment features already built-in to their cards.

Profitable deployments have usually been driven by a clear desire to gain market share and provide differentiating services that simplify the customer's life, while less profitable deployments have almost always turned out to be projects driven by a desire to test the complex technical capabilities of smart cards. The high cost of migrating to EMV is causing bank executives to pressure their implementation teams to choose the lowest-cost card and terminal hardware while at the same time exploring ways to maximize benefits for customers and make their cards stand out from all other payment methods, whether cash, cheque or other cards in the customer's wallet.

Not all European payment card markets are the same. There are major differences between countries in terms of competition and market saturation. Figure 1.2 shows the relative credit and debit card markets across major European countries.

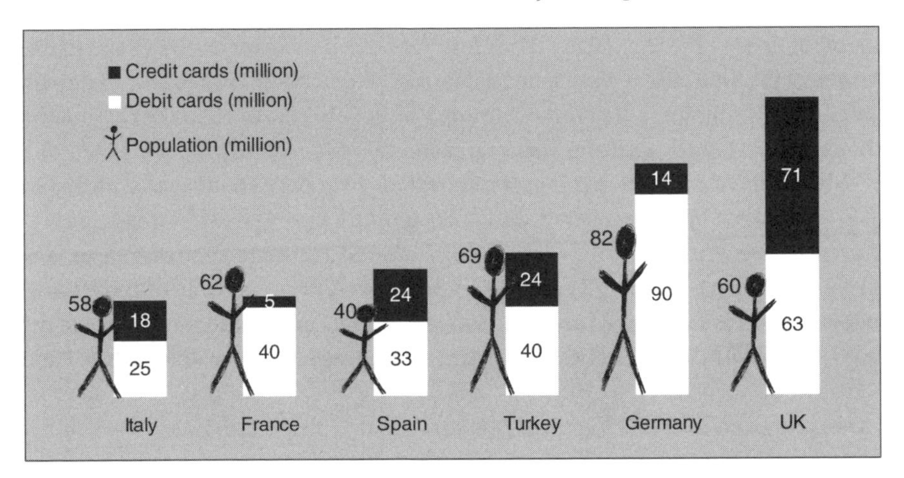

Figure 1.2 European bank card markets: number of cards in each country

The UK market is by far the most advanced. It is number one in credit cards and number two in debit cards. Germany is number one in debit cards but has a poorly developed credit card market. Turkey's debit card market is approximately equivalent to France's, but Turkey enjoys a far more mature credit card market. The country already has 24 million credit and 40 million debit cards and now ranks third in overall card and transaction numbers in Europe. The Turkish credit card market has grown quickly over the past three years, due to fierce and dynamic competition between the top two credit card issuers, Akbank and Garanti. We will explore the competitive dynamics between these two banks in greater detail. As credit card markets in Germany, France and Italy develop, bankers in these countries may run into some of the same issues faced by bankers in Turkey.

ATTRACTING NEW CARDHOLDERS AND MERCHANTS

In 2001, a European bank in Turkey, Akbank, created a new credit card, called 'Axess', which was launched as the first chip-based credit card combining instant surprise rewards and cash-back features. Thanks to this unique combination of card features, Axess gained one million new customers in its first eight months and won an award from MasterCard as Europe's fastest-growing credit card. The cash-back feature returns real value to cardholders by giving back a percentage of all cumulative purchases and storing the cash back in the chip for easy redemption. Cash back is available instantly as currency for purchases at participating merchants. In addition to cash back, the customer also gets a chance of winning surprise gifts whenever the card is used. These surprises are in fact promotions offered by merchants as a way of automating their 'buy *x* get 1 free' type offers, delivering earned rewards instantly at the point of sale. Cardholders experience greater convenience because everything is on a single card. When paying with their Axess card, customers do not need to carry or remember to use different loyalty cards for every merchant, nor do they need to worry about missing valuable rewards and discounts. Since launching Axess, Akbank has experienced a considerable marketing advantage in terms of customer acquisition. Akbank's credit card portfolio has increased by 45% and the bank's share of the Turkish credit card market has increased by 53%. In 2003, Akbank was awarded Best Bank in Turkey by Euromoney and *The Banker*, for Akbank's investments in information technologies, new technology offered through the Axess card, and a rapidly rising market share. By 2004, Akbank ranked as the most profitable bank in Turkey. Akbank is the largest privately owned bank in Turkey with total assets of more than US$20 billion, 618 branches and of the most extensive service networks in Turkey.

One year prior to launching Axess, Akbank's main competitor, Garanti, had already successfully launched a coalition card, called Bonus, which was chip-based and offered a number of customer features like cash back. In its first year, Garanti's Bonus card had already attracted over a million cardholders, primarily by converting their existing cardholders over to the new coalition card.

Akbank's strategic goals to increase their growth potential were to:

- attract new customers and develop a broad-based customer portfolio;
- increase high-yield potential from retail banking customers;
- decrease operating costs;
- increase customer satisfaction;
- develop strategic partnerships and diversified products that take into account individual and lifestyle needs.

Thanks to Axess, these goals were all achieved quickly, resulting in substantial financial benefits for the bank.

According to Akbank General Manager Zafer Kurtul:

> The most important factor behind Axess Card's success was Akbank's customer-centred product design. Akbank is able to offer customers the opportunity to earn as they spend thanks to the advanced technology we use. With Axess, the consumer is offered advantages like the chance of winning surprise gifts, the concept of 'chip money', which can be spent as soon as it is earned, and special promotions, all of which make Axess a far more advanced product than its rivals.

The cardholder base continues to grow rapidly (see Figure 1.3) at a rate of 42% in 2003. Akbank's credit card business is growing almost twice as fast as the rest of the market. It's market share, in terms of number of cards, grew from 6.3% in 2001 to 12% in 2003.

Akbank was also able to substantially grow its merchant network, demonstrating that its credit card appeals to both customers and merchants. In 2003, the Axess merchant network grew to 46 000 merchants, four times the number in 2002, and now includes major retailers like Carrefour, Migros, BP and McDonald's. By the end of 2004, the network had grown to over 80 000 merchants.

'Our success is directly attributable to Axess as the first credit card developed primarily to address consumer needs and expectations,' says Zafer Kurtul.

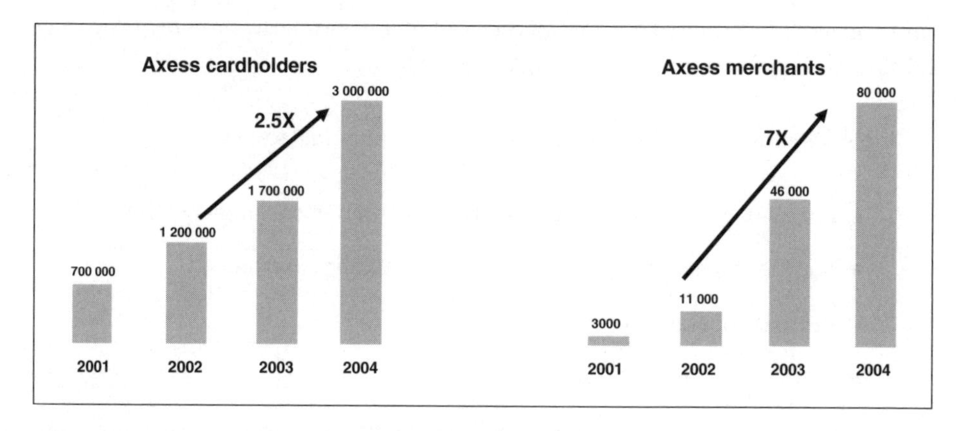

Figure 1.3 Growth in cardholders and merchants

Growth in Axess cardholders and merchants illustrates that once a bank reaches critical mass in terms of the number of customers carrying EMV payment cards, then merchant recruitment becomes much easier. This can be seen in Akbank's figures for 2003. When the customer base became large enough, merchants felt a greater need to not only accept the card for payment, but also to offer additional cash back and surprise gifts to customers. This created strong demand for Akbank's merchant acquiring activities. With greater demand due to differentiation, there is much less dependence on price. Other acquirers in other regions are beginning to show similar results. Some have indicated that EMV payment features have helped them boost their merchant base, while at the same time maintaining their merchant fees, and in some cases have even helped them increase fees with certain merchants who are beginning to use targeted promotions more intensively than others. Although Axess was launched one year after Garanti's Bonus card, within two years the two programmes were essentially tied in numbers of cards and participating merchants and by the end of 2004, the Axess merchant base had surpassed that of Bonus. Together, Axess and Bonus represent the lion's share of Turkish credit card growth from 2000 to 2004. Both banks successfully deployed chip technology five years prior to the Visa and MasterCard mandates, and both banks gained substantial financial benefit from migrating earlier than the other Turkish banks.

BOOSTING MARKET SHARE OF PROFITABLE CUSTOMERS

It is relatively easy to boost cardholder numbers substantially if the card is given away for free to anyone and everyone without paying attention to creditworthiness, which could result in lots of unprofitable customers. Akbank's annual financials are able to show that this was not the case.

The Axess customer base grew rapidly in 2003, at a rate of 42%. If Akbank were recruiting customers within an average category of creditworthiness, or consumers who want to take on new credit at a rate equivalent to the overall national average, we would expect the bank's market share of consumer loans to grow by an equivalent 42%. If, on the other hand, the bank were recruiting customers that posed a significant credit risk, or customers not interested in borrowing from Akbank, we would then expect the bank's market share of consumer loans to grow more slowly than the cardholder base. But, in fact, Akbank's market share of consumer loans grew much faster, at a rate of 64%, indicating that Akbank has been successful in attracting high-yield customers (Figure 1.4).

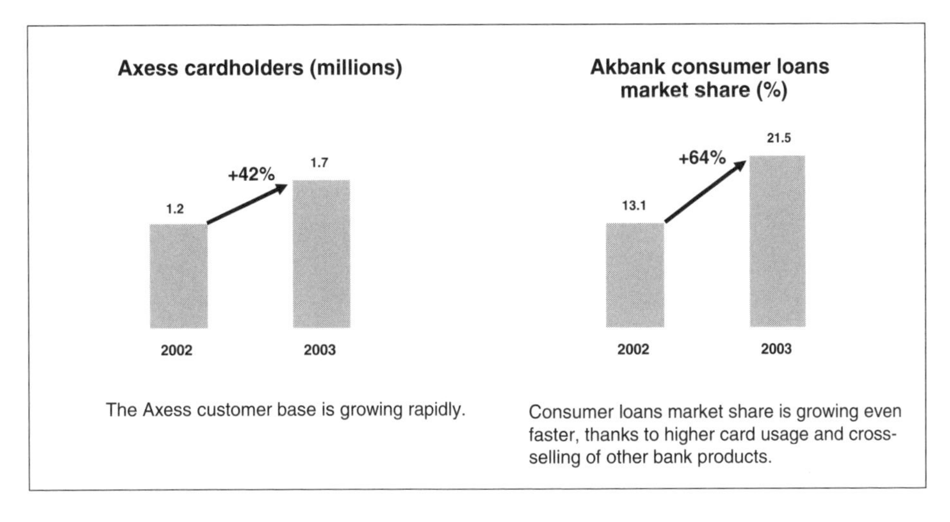

Figure 1.4 Growth in consumer loans market share

Fifty-nine per cent of Axess cardholders are new customers to the bank, and two out of three have become customers of other banking services. Akbank's market share of consumer loans has grown faster than the cardholder base thanks to higher card usage and, especially, thanks to successful cross-selling of other bank products to new customers. This is a key finding which we have seen in no other literature. It would appear that when a bank upgrades its payment systems and offers customers a new way to pay, new cardholders are more open to the bank's other products and are more likely to examine the possibility of taking out additional loans. The conclusions from a marketing perspective are clear. When you migrate to EMV, don't take the risk of wasting a special opportunity to cross-sell other products and services. Otherwise, your competitor may end up winning lucrative business from your customers instead.

BOOSTING INCOME FROM CARD FEES AND COMMISSIONS

Income from fees and commissions has benefited from the substantial growth in Axess card customers and transactions. Total net fees and commissions grew 57% in 2003, reaching US$151 million. Growth was driven primarily by credit cards, which now generate more than half of total fees and commissions (Figure 1.5).

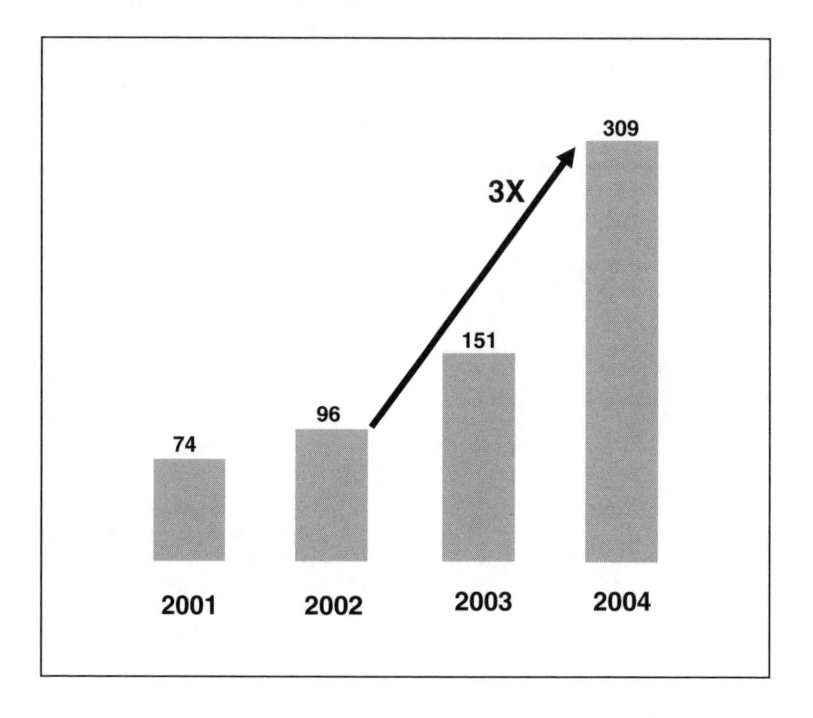

Figure 1.5 Growth in fees and commissions revenue (US$ m)

It is interesting to note that, as an added bonus for the cardholder, there is no annual fee for an Axess card, including for additional Axess card members on the same account. Fee growth came entirely out of other card fee categories, such as transaction fees, settlement fees and so on.

When EMV is used to offer new features which attract cardholders and merchants, payment processing suddenly appears to be less of a commodity. When a bank offers its customers what appears to be a completely new way to pay, very different from other payment card offerings, there is less pressure to reduce fees. EMV migration is an excellent opportunity for banks to create new payment features and services which can command higher fees and commissions.

We saw earlier how merchant associations provide their members with information to combat credit and debit card fees. Rather than stressing the point that fees charged by card companies on a gallon of gasoline are often greater than the profits earned by the retailer selling the gasoline, imagine how different the website of the National Association of Convenience Stores could be if the association saw plastic payment as a strategic part of its store operations, rather than an expensive commodity. Imagine the same website comparing card fees to the total amount spent on advertising, marketing, promotional activities, couponing, discounts and trial offers. Imagine the website comparing payment services to other IT services which can help merchants understand customer behaviour better, but which are far more cumbersome and expensive. It is very complicated today for a convenience store to recognize a new customer, a customer who has shopped frequently over the past month, or one who used to shop frequently but has not bought anything recently. Discounts and other promotional budgets are rarely targeted to specific buyers based on their purchasing behaviour, simply because the existing methods are all so cumbersome and expensive. The objective is to use EMV to create a new way to pay that helps merchants target their customers better, without requiring customers or merchants to do anything different at the moment of purchase, and for a cost that is already built into the merchant's standard payment infrastructure cost.

REDUCING THE COST OF ACQUIRING NEW CARDHOLDERS

Akbank acquired almost twice as many cardholders as it originally anticipated, without substantially increasing its advertising and marketing budgets. Prior to launch, Akbank established a marketing and advertising budget to acquire approximately half a million new cardholders within the first year. Without substantially increasing that budget, the bank acquired one million new cardholders in only eight months.

The average cost of acquiring a new cardholder is often estimated at over US$120 in many credit card markets, with higher figures in the most mature markets such as the US and the UK (see Figure 1.6). In this example, for a portfolio of one million cards, the estimated savings were valued at between US$30 million and US$60 million.

The trend in response to credit card solicitation mailshots in highly competitive markets has fallen below 0.6%. Recruiting cardholders through partnerships with merchants can be far more effective. The cost of cardholder acquisition can be dramatically reduced through relationships with merchant organizations. When the card is promoted at merchant outlets and clear benefits are linked to the use of the card at those outlets, a receptive audience immediately becomes available. Case studies exist where the bank pays US$10 to the merchant for each approved application form distributed through their network which effectively reduces cardholder acquisition costs to a fraction of their normal costs.

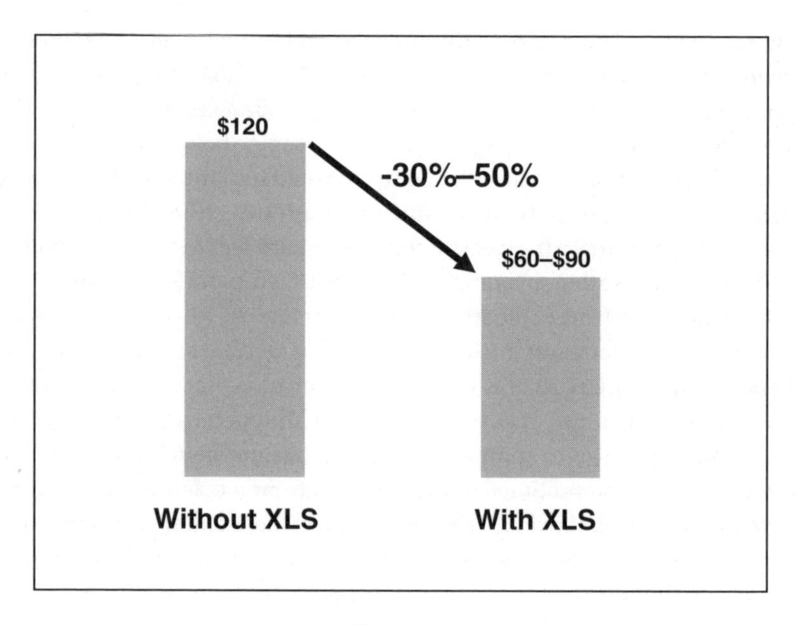

Figure 1.6 Estimated average cardholder acquisition costs (US$ m)

REDUCING MARKETING AND OPERATING COSTS

Cash back stored in the chip lets customers enjoy their rewards whenever they want and it eliminates the cost and hassle of traditional rewards programmes. Cash back, or points, is just another way for customers to pay for goods and services.

In the past, marketing budgets were largely spent on logistics, such as catalogue printing and mailing, statement generation, call centre support for balance inquiries, rewards inventory management and rewards request-processing and delivery to customers. Smart cards help eliminate most of these costs through real-time processing of rewards at the point of sale. An added benefit is that customers have more choice: they are not limited to selecting items out of a rewards catalogue, but can use their cash back or points to pay for anything sold at participating merchant outlets.

With EMV, a rewards statement can be provided on each credit card receipt, whereas traditional programmes rely on statements sent by mail. Not only is the cost of communicating information to the cardholder dramatically reduced, it is conveyed at a time when they are responsive to messages. The messages are therefore more likely to directly influence a cardholder's purchasing behaviour.

With traditional programmes, call centres are frequently faced with a significant increase in balance inquiries immediately after statements are sent out, because recent transactions often don't appear on statements. As soon as statements are sent out, call centres receive a number of calls from customers asking why the points or cash back related to a recent purchase did not appear on the statement. The call centre operator has to look up the account and will invariably find that the purchase is indeed recorded, but simply did not appear on the most recent statement. This represents a significant cost, as an average call centre or branch inquiry on a points balance typically costs in excess of US$2 to deliver.

When banks allocate 1% of credit card transactions to a catalogue programme, almost half of the budget can be wasted on logistics, whereas with EMV, customers can convert their points, miles or cash back at any participating merchant. Waste is eliminated and customers can enjoy double the reward value at the same cost to the bank (see Figure 1.7).

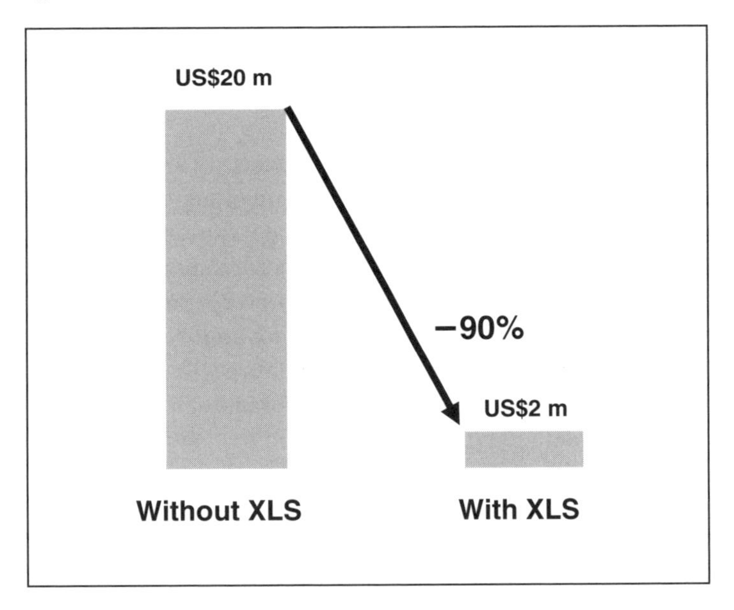

Figure 1.7 Estimated annual operating costs for one million active cardholders

Smart cards also help share the cost of rewards with other parties. Many banks encourage merchants to offer and pay for special bonuses, promotions and so on, linked to credit card payments. This is made easier through real-time delivery of special promotions that a merchant can target based on a customer's actual behaviour towards that merchant. Merchants already offer 'buy 1 and get 1 free'-type promotions on a regular basis. They offer discount coupons in the local paper, by mail, or distributed directly in their stores. But these coupons are rarely targeted, if ever. And they are expensive. Taking out a half-page advert in a single Sunday paper, for three or four coupons, costs between US$11 000 and US$14 000. Inserting a single-page ad in a co-op package mailed to a targeted neighbourhood costs between US$400 and US$500 for a single mailing. Using EMV, merchants can target their promotions better, by using behaviour data in the chip to cause the payment terminal to print out the coupons that the merchant normally gives out blindly. This can create significant discounts for cardholders, paid for by merchants.

REDUCING TRANSACTION PROCESSING COSTS

The French debit card system has become the world's most efficient, with transaction processing costs lower than in most other markets, since many transactions are handled offline and cleared and processed at night. The cards can even be used at parking meters and to pay highway tolls for very small amounts, often under a euro. They can be used at any McDonald's restaurant in France, and even at small fruit and vegetable stores. There are no receipts to sign: customers merely type in their PIN code and the whole process typically takes a few seconds, even counting the time it takes to punch the terminal's keys and print the receipt.

EMV was established to help reduce the dependence on the inter-banking network and dial-up connections to the network by providing the ability to conduct transactions offline from the bank. Depending on the card issuer's risk management policy, EMV can be configured to authorize a number of transactions between the card and terminal without going online to the bank for every transaction. EMV provides parameters within the card which establish limits for approving transactions offline, such as the number and cumulative value of consecutive offline transactions. Once the limits established by the bank are reached, the next transaction is forced online, where the bank can then update the EMV application in the card and collect the offline transaction data.

A general-purpose credit or debit card which can easily be used for low-value transactions, say transactions under $10, will always perform better than a dedicated e-purse that requires customers to load money to the card before using it and whenever it is empty. When compared to a dedicated low-value payment product – for example Belgium's Proton e-purse or France's Moneo – the general-purpose French debit card system proves to be more attractive to cardholders and merchants. Within two years of their launch, French debit cards generated 16 times as many transactions per cardholder as Belgian e-purse cards during Proton's comparable post-launch period and there were proportionally 4.5 times more places to use the card (one terminal for every 96 inhabitants for Carte Bancaire versus one terminal for every 435 inhabitants for Proton).

After spending substantial amounts of money testing various dedicated e-purse products in the late 1990s, both MasterCard and Visa have switched gears. Both associations have launched specifications for greater risk management for offline credit and debit transactions – MasterCard's M/Chip 4 and Visa's VSDC Plus – which can be used with general-purpose credit and debit cards. The specifications extend Visa and MasterCard's EMV debit and credit specifications to support pre-authorized spending limits for debit and credit cards.

According to Alexandre Cunesco, associate vice president and product manager, Global Chip Centre of Excellence, MasterCard International:

> In plain vanilla EMV, the card risk management features were designed to allow the issuers to implement the acceptable amount of offline risk they were ready to take. Now we are proposing a solution to make sure no offline risk will happen. We are hearing from the market that there are quite a number of merchant categories that are primarily cash-based today that would be willing to move to card-based transactions providing the cost of online authorization can be removed.[19]

He points at convenience stores and campus retailers. Cash substitution will also reduce the risk and cost of handling cash for merchants.

When all transactions need to be authorized online all the time, any increase in the number of transactions performed with payment cards will automatically result in an almost directly proportional increase in the cost of processing those transactions. Processing ten million more transactions each month requires almost ten times more back-end support, machines, processors, throughput, etc., than for one million transactions per month. There are certainly some economies of scale, and central processing unit (CPU) costs are certainly getting cheaper all the time, but in general any increase in transactions will result in something close to a proportional increase in processing costs.

On the other hand, when a growing number of transactions can be processed offline, with no additional demand on a bank's central authorization systems, the number of transactions can increase substantially with very little impact on processing costs. Transactions are uploaded at night and processed as a batch, which is more efficient and far less costly (see Figure 1.8).

Banks can use EMV to make their credit and debit cards more efficient, and win market share away from cash and cheques. Chip debit and credit cards can open new payment markets without the hassles and risks of creating a completely new paradigm like an e-purse card that has to be reloaded at a teller machine whenever it is empty. Banks and the major card associations have long been keen to move a significant percentage of the world's hundreds of millions of cash customers onto cards, and the key to doing this is to use offline payment schemes for low-value transactions. In the

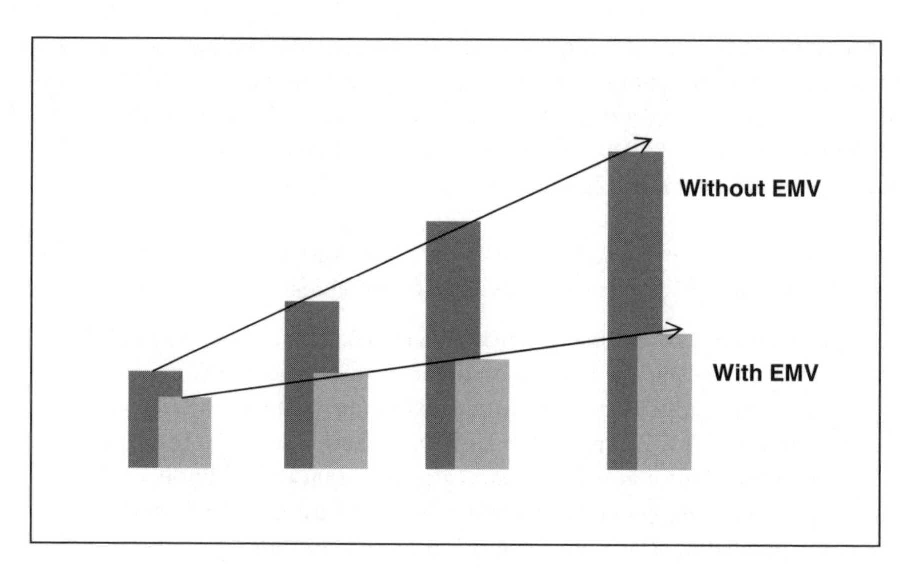

Figure 1.8 Growth in annual transaction processing costs, with and without EMV

past, the industry got caught up in cumbersome e-purse approaches to the problem. Today, the underlying elegance of card products like M/Chip 4 and VSDC Plus is that it's all EMV.

UNLOCKING THE VALUE HIDING IN YOUR EMV INVESTMENT

We have seen a number of projects recently in which, for every $100 spent on upgrading a bank's card, terminal, processes and systems and support infrastructure for EMV features designed to combat fraud, as little as an additional $2 to $5 is needed to add payment-enhancement features necessary to combat commoditization. This is possible when the bank's high-level mandate is to build market share at the lowest possible cost and to take advantage of EMV migration to create differentiation. Such a mandate creates a pragmatic mindset within the implementation team so that unnecessary investment in futuristic technology can be avoided.

According to Gartner Research, in most EMV deployments across the world, banks have invariably chosen the low-end single-application 4K EEPROM cards. Virtually all of these cards support all EMV features, the basic anti-fraud features as well as customer-centric enhancements that provide a powerful tool to fight commoditization. All of the major smart card and terminal manufacturers – Axalto, Gemplus, Giesecke & Devrient, Hypercom, Ingenico, Lipman, Oberthur, Sagem, Verifone – offer full-featured EMV products on a wide range of hardware platforms and low-cost single-application EMV cards as well as more expensive multi-application Java and Multos cards. Banks can benefit without choosing special hardware or switching suppliers, and without buying more expensive cards or terminals. This keeps the cost of full-featured EMV virtually the same as a less complete EMV implementation limited to fraud prevention features only (Figure 1.9).

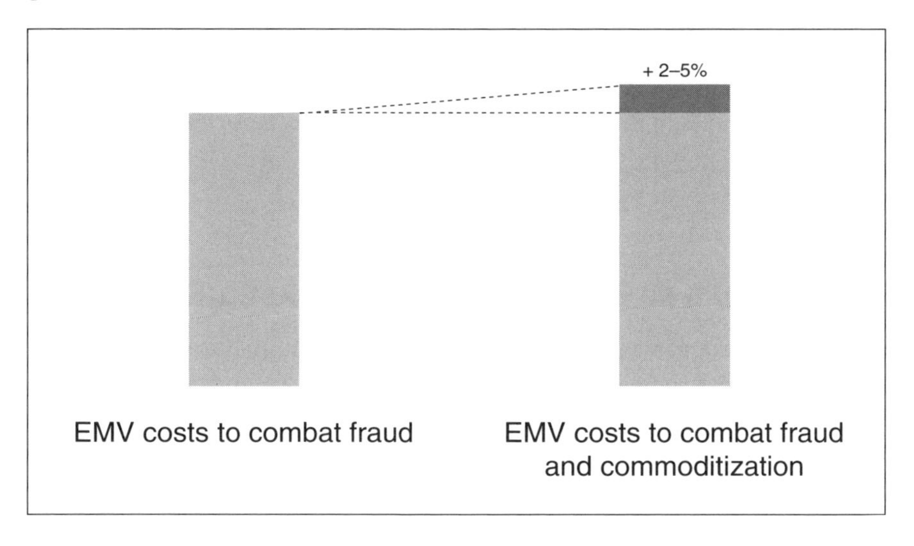

Figure 1.9 EMV costs to combat both fraud and commoditization

Since most cards and terminals on the market today support fully enhanced EMV, most banks are probably already about to deploy the necessary cards and terminals. They should be able to offer customers additional features without buying more expensive equipment.

EMV enhanced payment data structures need to be added to cards during the card personalization process, in the same way that EMV fraud management data is added. Because the data is added at personalization, not only does this eliminate the cost of more expensive multi-application cards, but it also saves on the cost of acquiring, installing and operating a multi-application card management system (or 'CMS') and avoids the complexity of managing multiple applets across each card's life-cycle.

How about operating the service? How many people are needed to set enhanced EMV parameters downloaded to merchant payment terminals? Since merchants can potentially each have their own special parameters and may change them frequently, one might imagine that a substantial number of specially trained people are required. In fact, operating a terminal management system, setting special marketing parameters, downloading them to terminals, monitoring the uploaded transaction details and so on requires little additional overhead on top of the services already provided by a merchant-acquirer. Akbank operates the service with two people, working part-time, for over 80 000 merchants.

A card with a $1 chip and 4K memory that only supports PIN verification is incomplete, as the same card can also support welcome gifts, surprise bonuses, cash back and loyalty points. Why risk investing in EMV cards and terminals without basic features that customers enjoy and that attract merchants? Why simply become EMV-compliant, just like all the other banks, when you can have the most attractive payment system in your region for virtually the same cost? These are important questions that should be asked before investing in EMV.

CASE STUDIES IN EMV MIGRATION STRATEGIES

In an Eastern European country, two banks are direct competitors: Edelweiss Bank and CTZ Bank (not their real names). Edelweiss Bank is the country's largest credit and debit card issuer, with around 40% market share. CTZ Bank's card base is around a third the size of Edelweiss's, but CTZ is by far the largest merchant-acquirer, with over 60% market share of merchant outlets. CTZ Bank's non-interest income increased by 52% in the most recent fiscal year, much of the growth due to merchant processing fees. In comparison, Edelweiss Bank's non-interest income increased by less than 5%.

Financial analysts have been encouraging Edelweiss Bank to become less dependent on interest income and work on growing non-interest income faster. Whereas analysts have been insisting that CTZ Bank must expand its cardholder base, they are nevertheless satisfied by CTZ Bank's rapid growth in non-interest income and the stability that those income streams will bring during difficult economic times.

In each bank's most recent annual report, both chairmen state that GDP growth has slowed, income from loans is stagnating, and share of non-interest income is growing but needs to grow faster. Both chairmen are confident that the growing credit card market will be lucrative.

Both banks plan to upgrade to EMV during the same year. Both intend to use the lowest-cost single application EMV smart cards available. Both have chosen cards which also support enhanced EMV features in addition to fraud features, to reduce the impact of commoditization at the same time that they are reducing losses due to fraud.

Some of the strategic questions which Edelweiss Bank needs to be addressing as it prepares to upgrade its payments infrastructure for EMV are:

- How can I use my large card base, in conjunction with EMV, to attract new merchants and increase my acquiring activities to become the leading merchant-acquirer so that I can dramatically increase payment fees and commissions?
- What might happen if my direct competitor, CTZ, offer enhanced EMV services first? Its large merchant base will be attractive to cardholders and may cause it to grow its cardholder base much faster than me, even drawing my customers away.
- How will analysts and shareholders react if my competition continues to increase its non-interest income and mine remains stagnant? If GDP grows faster than expected and if lending picks up again, no problem. But what if that doesn't happen?

Similarly, some of the strategic questions which CTZ Bank needs to be asking include the following:

- How can I use my large merchant base, in conjunction with EMV, to attract new cardholders and become the leading card issuer?
- What might happen if Edelweiss offers enhanced EMV services first? Its large cardholder base will be attractive to merchants and may cause it to grow its merchant base much faster than me, which may cause my non-interest income to stagnate, or worse, it may draw my merchants away.
- How will analysts and shareholders react if I cannot continue growing non-interest income at similar levels as the last two years? If I can continue growing my merchant base so that I can process a larger share of transactions than my competition, no problem. But what if growth is slower than expected? What if the competition begins taking my merchants away?

In another European country the top Visa card issuer is proud of having been the first to issue a very popular credit card that allowed customers to accumulate air miles. Today, the credit card market is mature and a number of other issuers also offer air miles to their customers. Each adult already has an average of over two cards and there is virtually no room for growth through new customers without cards. Competitors are naturally targeting the bank's customer base. Fraud is under control, but the bank intends to beat its competitors to market by being the first to deploy EMV cards with customer-centric enhancements. The bank intends to (a) leverage EMV to make the payment experience more exciting with new value-added services at the moment of purchase, to help boost card revenues and secure market share, (b) enhance its existing air miles card with new real-time features to differentiate against other similar products from competitors, and (c) establish a lasting competitive advantage so that the bank can secure greater benefit from EMV migration than other players and avoid being copied by competitors.

In January 2004, Mashreqbank, one of the oldest banks in the United Arab Emirates and the leading merchant-acquirer with 40% of the UAE acquiring market, launched the first fully enabled chip card in the Middle East offering customers a richer, enhanced payment experience, at the same time as upgrading all its merchants' point of sale (POS) terminals. The impact on non-interest income was immediate. In the first nine months of 2004, non-interest income grew 36%, far exceeding net interest income growth of 13%. At present 50% of Mashreqbank's income is from non-interest sources, a trend the CEO is happy to see because this income carries no credit risk for the bank.[20]

'It is particularly gratifying to see good growth in our non-interest income as this is a sign of our customer's appreciation for our value-added services,' said Mashreqbank's CEO, Abdul-Aziz Al-Ghurair in the bank's third-quarter 2004 press release. 'Our strategy of investing in cutting-edge technology has given us an insurmountable lead in providing the products and services the public wants.'

After launching its own card (called WOW!) Mashreqbank leveraged its payment infrastructure to develop a co-branded card with Virgin Megastore and MasterCard. Jean Claude Torbey, regional marketing director of Virgin Megastore, said that in addition to collecting points from all around the world, the VIP Mashreqbank credit card gives special privileges; these include invitations to concerts and movie previews, in-store promotions, a 0% instalment scheme for purchases from Virgin Megastore, loyalty-based discounts at over 500 outlets across the UAE and first-to-know updates from the entertainment industry.

'The response to the trendy, non-square credit card has been overwhelming,' said Mashreqbank's Steven Pinto. 'Niche marketing to specific demographic groups certainly pays off as it provides a clear point of differentiation from generic, one-size-fits-all credit card offerings.'[21]

Malaysia is the first country in the Asia-Pacific region to complete national migration to EMV. Initial decisions by three of the top five Malaysian banks already lead to predictions that perhaps 80% of EMV cards to be issued are likely to be enhanced with payment features including loyalty points and instant promotions. HSBC Malaysia, for example, have determined that implementation of EMV smart cards will not only help reduce card fraud but also provide a platform to make its cards more popular with its customers and will differentiate the company from its competitors. Likewise with Taishin bank, the second largest card-issuing bank in Taiwan, whose magnetic strip credit cards are due to be upgraded to chip. Taishin decided to take advantage of this migration and enhance the value of its EMV cards with additional payment features. Komerční Banka (KB), subsidiary of Société Générale, the top merchant-acquiring and leading card-issuing bank in the Czech Republic, has already begun converting all of its credit and debit cards to EMV and will enhance its cards with numerous added-value services, such as gifts, loyalty points and surprise bonuses, designed to appeal to a wide range of customers and merchant segments. Similar trends are evident in other parts of the world, wherever the move to EMV is accelerating and wherever competition between card issuers and merchant-acquirers is becoming more and more difficult.

EMV migration is not just a technical upgrade to a bank's payments infrastructure. Banks could risk losing a lot if they concentrate on EMV solely as a technical method to reduce fraud and invest heavily to do no more than implement fraud reduction features. EMV needs to be seen as a unique opportunity for a bank to also upgrade its competitive strategies and resist commoditization. EMV migration can provide the foundation for growth opportunities in banking, but only if banks look at EMV in a wider context than simply fraud reduction, and concentrate equally on EMV features designed to increase revenues, market share and profitability.

THE THREATS: COMMODITIZATION AND FRAUD – FOUR KEY OBSERVATIONS

1. UK authorities may force banks to reduce transaction fees charged to merchants, resulting in a loss of revenues estimated at £1 billion per year, more than twice the losses due to fraud. In Australia, the Reserve Bank has reduced annual bank revenues by AU$430 million, more than four times the level of fraud. In the US, the Visa and MasterCard anti-trust settlement could cause banks to lose fee revenue of over US$10 billion per year, between three and four time the losses due to fraud.

2. Reduced fees are only one part of the commoditization problem. UK banks are also losing another £1 billion per year due to customers switching card companies, seeing no real difference between cards other than the rates. Commoditization costs UK banks at least £2 billion per year, a problem five times as great as fraud. Commoditization simply dwarfs fraud.

3. As banks discover that they can use EMV to address both fraud and commoditization, as opposed to fraud alone, for a virtually insignificant increase in the total cost of EMV migration, more and more of them are choosing to do precisely that. Full-featured EMV, including customer-centric enhancements, can provide valuable differentiation in an increasingly competitive market, with little impact on the total cost of EMV migration. An EMV infrastructure which includes full enhancements, in

addition to basic fraud features, costs little more than basic EMV, but offers far more benefits for cardholders, merchants and bankers.

4. A card with a $1 chip and 4K memory that only supports PIN verification is incomplete, as the same card can also support welcome gifts, surprise bonuses, cash back and loyalty points. Why risk investing in EMV cards and terminals without basic features that customers enjoy and that attract merchants? Why simply become EMV-compliant, just like all the other banks, when you can have the most attractive payment system in your region for virtually the same cost?

CHAPTER 2

Creating Competitive Advantage through a Richer Payment Experience

The customer experience is the next competitive battleground.

Jerry Gregoire, former chief information officer, Dell[22]

I am standing in line at a checkout counter, about to pay. The experience I am about to have will depend on several things. First of all, the enthusiasm and cheerfulness of the store representative standing in front of me. This will have a big impact on the quality of my experience. Then, as the last step of the process, the payment method I choose to use. My experience at the moment of purchase will be slightly different depending on whether I pay by cash, cheque or plastic. With cash, I will need to deal with change. If I pay by cheque, I will probably need to show some form of identification. If I pay with plastic, I will need to sign the receipt or key in my PIN code, depending on which piece of plastic I pull out. That's about the full extent of my payment experience. At the moment of purchase, there is really very little difference between these payment methods. Other things may happen later, concerning account balances, mileage points, and other such things of which I will be notified by mail. But currently that's about all there is to it.

The point of purchase has become a critical part of ensuring a pleasant overall shopping experience. In many retail segments, checkout hardware components have gone from being a necessary expense to being a business driver. Retailers need every advantage they can get to differentiate themselves from their competition, and they are using technology at the checkout to do that. There are many things that retailers are already doing to make a customer's experience more enjoyable, whether it's a faster checkout process, the ability to check inventory in other stores, providing gift receipts, taking customer data to customize coupon offerings, or making special offers. There are many ways of using point of sale hardware equipment to make a difference in a customer's experience.

Of course, all of that investment is useless if front-line employees don't show enough enthusiasm, charm and knowledge. An engineering study conducted by IBM's Advanced Business Institute demonstrated that the payment experience had a significantly higher positive impact on the perceptions of departing customers when the service employee looked them in the eye. To conduct the study, IBM added a few dummy keys to the cash register and modified the checkout software so that a customer's receipt didn't print until the dummy key labelled with the colour of the customer's eyes was pressed. To get the receipt to print, the clerk had to look the customer in the eye, then press the corresponding key.[23] This confirms what all customers already know – connecting with the customer in a positive fashion for even a very brief moment makes a world of difference.

In very general terms, the payment experience has not kept up with the advances that retailers have achieved with the rest of the checkout process. Credit and debit card payment technology providers have focused exclusively on making the transaction as fast and as secure as possible. Now that all card products have essentially reached the same level of speed and security, banks are beginning to focus on other important elements which contribute to a customer's pleasant payment experience. The moment of purchase is one of the very few moments when a customer is holding your card in their hands, touching it, getting ready to use it. The rest of the time it's in the customer's wallet and probably out of their mind. A remarkable opportunity exists to provide an extraordinary customer experience at the moment of purchase, and become truly differentiated in the eyes of your customer. Customer experience is much more than price, quality or even customer service. When you look closely, you will see that in 'best of class' companies, the customer experience is a fundamental component of the business strategy. The experience a customer enjoys while using your card to pay will become the next competitive battleground. And EMV is poised to play a major role.

Here are some of the things that my chip-equipped payment card can do for me at a merchant's point of sale terminal:

- The merchant offers me a welcome gift the first time I use my card there. All I did was give my credit card to the clerk, who gave it back to me with my receipt, along with the gift. Then I notice that the bottom of the receipt says 'Good for 1 welcome gift'. The process didn't take any longer than normal.
- I've been having lunch frequently at a restaurant near my office. Today, surprise, my receipt tells me I'm entitled to a free dessert at my next visit.
- After work, when I buy a new DVD, the credit card receipt reminds me that my wife's birthday is coming up soon. Of course, I didn't forget. But since I'm out shopping anyway I decide to pick something up before I do forget.
- I see my points balance at the bottom of every receipt, so I always know how many points I have.

- I can choose to pay with my points, rather than credit, without going through the tedious process of filling out redemption forms or selecting rewards from catalogues, which almost always lack items that I am really interested in. I pay for a second DVD using the points in my card. All I had to do was tell the clerk that I want to pay with points, rather than my card's credit facility.
- I receive a $10 gift certificate after having significantly increased my purchases at a particular store.
- At my first visit to a sports store that I don't normally go to, I'm apparently recognized as potentially a very valuable customer – I instantly receive a gift certificate saying so. Then I remember that when I applied for my new card I said that I like sports.
- When I go back to a store that I haven't been to in a long time, but that I used to shop at regularly, my credit card receipt tells me that I'm entitled to a bottle of champagne – and invites me to come back here more often. Don't be a stranger, it says on the receipt. I hope the champagne is good.

If you are a card issuer, your EMV payment infrastructure can help you enhance the payment capabilities of your card by adding customer-centric features whenever your card is used to pay. The payment receipt is an ideal communications tool. You can add lots of useful information to the receipt, like a customer's cumulative points balance, or special marketing messages encouraging the customer to use their card more frequently. Customers can redeem their points directly at the moment of purchase, using the same payment terminal used for a credit or debit transaction. How much can you save by eliminating a rewards catalogue, warehouse and redemption process?

If you are an acquirer or a payment processor, your EMV payment infrastructure can help merchants know that a customer is using their card in that merchant's store or chain for the first time and can inform them when a customer's behaviour has just changed dramatically. How useful would it be for a merchant to know that the customer standing in front of her has just doubled his monthly spending and that his projected lifetime value is now just over $500 000? She might want to offer that customer a special gift, a bottle of wine or a ticket to the opera, discreetly, just by printing the offer at the bottom of the credit card receipt. The customer can pocket the receipt, use it later, or ignore the offer and throw it away. It is mechanical, automatic and non-intrusive.

On a more defensive note, if a customer has reduced his spending recently, the merchant might guess that he is probably spending more at the competitor's store. Since he's standing right in front of the merchant now, maybe something can be done about it.

What if a new customer begins shopping at the merchant's store and you can instantly provide the merchant with a way of knowing that her customer currently spends a lot in that merchant's retail category? Wouldn't she want to look for ways to treat this brand new customer with great care, even if he doesn't currently spend much in her store?

All of this information can be available through your EMV infrastructure. Provide it to your merchant customers for a fee, on top of what you normally charge for processing payment transactions. Plus, when merchants link their incentives to your EMV infrastructure, they will actively promote your payment brand, for example by placing in-store signs offering a free gift the fourth time customers pay with their card, or after the customer spends $100. When a merchant's marketing programme is linked to your cards, you can bet the merchant will do everything possible to help customers become cardholders by placing take-ones directly at the store counter and systematically asking for the card. You can also be certain that merchants will go out of their way to make sure the cards are actively used.

Here is the real-time message, according to Regis McKenna:[24]

> New consumers are never satisfied consumers. Managers hoping to serve them must work to eliminate time and space constraints on service. They must push the technological bandwidth with interactive dialogue systems – equipped with advanced software interfaces – in the interest of forging more intimate ties with these consumers. Managers must exploit every available means to obtain their end: building self-satisfaction capabilities into services and products and providing customers with access anytime, anywhere.

SURPRISE GIFTS, DISCOUNTS AND PRIVILEGES: MAKING THE PAYMENT EXPERIENCE MORE FUN AND EXCITING

Many, if not most, merchants already offer paper incentives to their customers, such as coupons distributed in the Sunday paper or printed in leaflets stuffed into letterboxes. Rather than distribute these incentives blindly, merchants can use your EMV infrastructure to deliver their highest value promotions to their best customers. EMV cards that keep track of recent customer behaviour can also help to streamline the way soft benefits are managed, benefits that are not necessarily promotional discounts such as a welcome gift, VIP access, special upgrades or advance notice of sales. Or even a simple thank-you note to best customers, printed at the bottom of the credit card receipt, with no special offer attached. Soft benefits appeal to the customer's emotions, aspirations and pleasure in being recognized. They tend to be more powerful and 'sticky' than hard, financial benefits.

Virtually all EMV cards available today are capable of keeping track of the customer's prior purchasing behaviour across many different stores. This gives merchants a simple way of providing perks to their best customers. After a certain number of visits or amount of money spent at a particular store or chain – information stored in the card's chip – the POS terminal prints out an incentive message at the bottom of the receipt, or flashes it on the POS terminal display. Using behaviour data stored in an EMV card to trigger welcome gifts, VIP access privileges, coupons, discounts and other incentives provides significant value to the merchant, the cardholder and the card issuer. When merchants use your payment infrastructure to offer their own targeted promotions, the value of incentives provided to your cardholders is much greater, while at the same time reducing your operational responsibilities, since redemption and clearing of points across numerous merchants is not required for these promotions. If your goal is to make your card programme simple, scalable, inexpensive and easy to install, deploy and operate, make sure that your EMV infrastructure has these capabilities built in from the start.

Merchants know how to reward their best customers and don't need much marketing help in defining the incentives they use today. Many widely used paper methods have been around for a long time and work well. If you are an acquirer, ask one of your merchant customers if she knows which coupons work the best for her store and generate the greatest customer interest. She will instantly respond, perhaps suggesting something like a 2-for-1 meal, in the case of a restaurant for example. Then ask her if she gives those offers out frequently. Since a 2-for-1 offer costs a lot, she most likely saves them for special competitive situations when there is a need to boost sales. Then ask if she would like to be able to print that 2-for-1 offer at the bottom of card receipts for customers that have purchased a lot recently or customers that have not been to her restaurant in quite some time. Also ask if she would like to be able to offer a welcome gift to customers who use their card for the first time in her restaurant. If she serves breakfast and lunch, perhaps she would like to use the credit card receipt to offer a breakfast discount to lunchtime customers. These discussions between an acquirer and the acquirer's merchant customer do not require specialized promotional marketing expertise – and they are much more productive discussions than haggling over the fees you charge.

Some merchants go a step beyond coupons and offer their customers a small card that is punched or stamped at each visit. After a certain number of visits, the customer gets something free. These merchants will immediately understand the concept of triggering a free item based on the number of transactions stored in the EMV card's chip. All the acquirer needs to do is set the parameters that are downloaded to the merchant's payment terminal.

How about being able to use your credit or debit card to jump to the head of the queue at movie theatres, sporting events, museums or concerts? Banks already negotiate VIP passes for their gold card customers, but the process of obtaining them is tedious. With an enhanced EMV card, customers could be eligible for the queue-jump service whenever they spend over a specified amount – say $1000 – in a single month: a VIP pass privilege that customers earn. In many cases, the venue already has special fast lanes which already have payment terminals. The terminals in those lanes simply need to be configured to recognize valid queue-jump customers. Another option is for the bank to provide key ring card readers to venue personnel, with a display that says whether or not the customer is entitled to VIP privileges. The privileges would remain in effect indefinitely, on condition that the customer continues to spend over $1000 each month.

French oil company Total was probably the first to issue a loyalty card in the early 1990s that offered a neat twist in addition to the usual one point for every ten litres of petrol purchased. If your car breaks down, for any reason at all, within 15 days after filling up at Total, on-the-spot assistance would be provided by Total. The advertising campaign was clear and simple. The message was well received, especially among Total's female customers, a primary target for the company's loyalty card programme. Fifteen days is a normal period of time between two fill-ups, so the motivation to go back to Total is strong. Why let an insurance policy expire when all you have to do is go back to Total for gas? There was a drawback, however. Customers had to pay for the towing services out of their pocket, keep the receipt, fill out a form, send everything by mail and wait for Total to process the request and send back a cheque. It would be easier if the garage mechanic only had to place the customer's enhanced EMV payment card in a reader, verify that assistance was paid for by Total, and be done with it.

Consumers today are bombarded with many different types of promotions. Free samples are among the most effective at generating trial and repeat purchases. Since the sample is free and is presented to the customer in person, the sample is sometimes seen as a gift. In some retail categories, like cosmetics, especially high-end brands, samples are more advantageous than discount coupons precisely because they are seen as a gift, a soft benefit, rather than a financial deal. Today, if you shop at many different types of retail store, the checkout assistant is likely to offer you a free sample, adding it to your bag. These samples are not targeted. You may get the same sample twice in a row. With an enhanced EMV payment system, the terminal can display 'Give sample pack 1' at the customer's first visit to a store, 'Give sample pack 2' at the second visit, and so on. The merchant can then be certain that customers are receiving samples of the full range of products available. The most expensive samples could then be limited to customers who have been to the store several times, further enhancing the idea that the sample is a gift, and that the customer is enjoying a special privilege of receiving something that many others do not receive.

The feature consists of a few simple parameters downloaded to the merchant's terminals. If the merchant has several stores, the same parameters can be sent to all the stores. Since the customer's behaviour data is stored in the card, the customer might receive her first sample gift in one store, and the second sample in another store. There is no need for the bank to be an expert in promotional marketing. All the acquirer needs to do is inform the merchant of the types of parameter which can be used to trigger any type of message that the merchant would like. It is a simple, enhanced payment feature which can be very useful in making the payment experience more fun, exciting and memorable.

Invite merchants to offer birthday promotions, for example a 2-for-1 meal in the case of a restaurant. Again, this could be little more than a special marketing message printed at the bottom of the receipt, triggered using information stored in the card. A merchant can choose to have the same message printed at the bottom of the receipt for all customers with a birthday coming up, or different messages can be provided based on how often the customer has shopped recently. Invite the restaurant chain to offer a free dessert to casual customers on their birthdays, and a bottle of wine from the restaurant's finest selection of wines to customers who already dine there frequently.

ENHANCING THE PAYMENT EXPERIENCE WITH REAL-TIME POINTS, MILEAGE AND CASH BACK

When one mentions loyalty, many people immediately think of accumulating value over time, saving up a small percentage of each transaction, as points or miles like those earned through frequent flier programmes. When the value counter is identical to the local currency, the technique is sometimes known as cash back.

Points, mileage or cash back programmes offered by card issuers generally allow cardholders to acquire value across many different merchants, every time the card is used. Points are saved up then redeemed towards rewards from a catalogue provided by the issuer, while mileage is usually redeemed for plane tickets. Cash back is simply deducted from the customer's statement, often at a fixed interval, such as once a year. Since the accumulated value represents real money to the issuer, these programmes are equivalent to providing a deferred discount to customers. Card issuers cannot give substantial discounts on their own, so most have entered into arrangements with retailers who can add their own discounts on top, for example by offering customers double or triple points when they use their card at selected merchants. When mileage points are given to a cardholder, the merchant immediately owes the card issuer the value of those points. If, for example, a merchant gives points at the rate of 2% of each transaction's purchase value, this is really just another way of giving customers a 2% discount. Rather than providing it directly to the customer, the merchant in essence gives the discount to the card issuer who manages it for the customer, almost always in exchange for a commission which is usually hidden in the cost of operating the system.

Virtually all credit cards, including private and co-branded cards, offer programmes based on mileage points. Today, these programmes are so similar that the web literally crawls with tables comparing dozens of credit card features such as where you can use them for mileage, how much each mile is worth, how many miles are required for a minimum reward and so on.

If the card issuer is capable of adding significant value to the accumulated points, the programme makes sense for everyone. The airline industry created the mileage concept in 1981, when American Airlines launched the AAdvantage frequent flier programme. Miles are good for free tickets and upgrades, which have a high perceived value, but represent a low incremental cost to the airline. Airline seats are 'perishable' – once the plane takes off, an empty seat can never be sold again. Filling an empty seat with a passenger that pays with miles costs little in incremental costs (US$18 to US$54, mostly for food and fuel, according to a Consumer Reports Study[25]) and can even generate additional revenue when the passenger travels with a spouse paying full fare. The wide gap between the perceived value of miles and their actual cost means that customers can acquire attractive benefits in a reasonable time frame. When the value of points is substantially greater than their cost, the chance for success is high. The economics described here are applicable to all industries that sell perishable items like airline seats, hotel rooms and car rentals.

A wise friend once told me that happiness equals reality divided by expectations. Personally, this has always made me smile. But the idea applies well here. 'Success equals mileage value divided by mileage cost.'

Successful frequent flier programmes have been copied by companies in many other industries, but often with less success. Non-perishable items have a real cost to the seller that is much closer to the customer's perceived cost than to an incremental cost approaching zero. Compared to airline seats, the retailer's cost of a box of cereal is very close to the customer's price. The result is that it takes a long time to acquire attractive benefits with supermarket and gas station loyalty programmes. When the cost of points is only slightly less than their value, it is much harder to make the programme work correctly from a financial perspective. Mileage programmes in these industries are forced to find a way to create a wider gap between the actual cost of the reward and its perceived value. This is difficult. Card issuers are generally faced with two options: either add meaningful value to the points, or dramatically reduce the cost of the rewards without sacrificing quality.

In too many cases the card issuer merely acts as a clearing mechanism, acquiring points from merchants that issue them to cardholders, and reimbursing merchants for points used for payment in their stores. It can be argued that, from a merchant's point of view, mileage is a complicated discount mechanism that doesn't necessarily build loyalty to their store. For customers, mileage is essentially a deferred discount that must be accumulated over a long period of time in order to be eligible for anything worthwhile.

Not only do mileage programmes typically fail to add value to the points they manage, but, worse, they sometimes even succeed in destroying value through high operating costs. When discounts offered by merchants are converted to mileage points managed in a central computer, costs like monthly statement mailings and voucher processing eat up a large portion of the discounts that were originally supposed to be for the customers' benefit. This is a problem that plagues mileage programmes in all industries. It becomes particularly unmanageable in industries that don't enjoy a high margin between the perceived value and the actual cost of mileage points.

If you are currently operating a mileage programme, here is a simple way to find out whether you add sufficient value to the points or not: ask merchants if they would prefer that the points they issue be converted to gift certificates valid only in their stores. In other words, would they prefer that the discounts they offer be used by customers to purchase additional products and services in their stores? If the answer is yes, you have a clear indication that merchants feel that your programme does not add value to the discounts they can directly provide to their customers. If you cringe at the idea of even asking this question of your merchant partners, well, you have your answer. A common complaint heard among merchants participating in poorly developed mileage programmes is that those programmes are a loyalty mechanism for the card issuer, not for the merchant. Since the card issuer needs the merchant's participation, and the merchant knows that their participation benefits the issuer, the merchant is in the driver's seat.

There are a few special situations in which a card issuer is able to add substantial value to collected points. When a card issuer partners with clubs, affiliations, charities and other cause-related organizations, such organizations can use the mileage points for specific causes that customers hold dear, the emotional element adding significant value to the mileage points. Cause-related mileage programmes are complementary to surprise gifts, discounts, privileges and other merchant promotions triggered using EMV behaviour data, discussed earlier. Targeted merchant promotions address the needs of all merchants and cardholders, while cause-related mileage points can be used to address the needs of specific segments of customers that would like to participate in cause marketing programmes, in addition to benefiting from merchant discounts.

Children's Heroes is a cause marketing concept in the US that uses merchant rebates to raise funds for schools. Chris Hutcherson, founder of Children's Heroes, was first to pioneer a paper gift certificate version of the programme beginning in 1991 under the industry name of Scrip. Over a period of six years, the programme went from one state to 44, from 385 schools to 11 000, from 40 000 families to four million, from eight major retailers to 265. From US$24 million in sales, generating US$1.4 million for schools, the programme reached US$1.6 billion in sales and generated US$107 million for schools.

Chris Hutcherson says that parents see participating retailers as:

> Corporate citizens and community leaders that care about and do good for our kids, our families and our schools. The love parents have for their children is the single most powerful force on earth. Our children stand at the centre of our universe. To nurture one's child is each parent's duty, but the education of every child transcends boundaries required of parents only. In the truest sense of duty to the communities in which we live, the well-being, the education and the future of the children on this planet is the responsibility of us all.

Cause-marketing concepts add substantial emotional value to points. Many parents would not hesitate when faced with the choice of pulling out a credit or debit card that earns a reward as well as money for their child's school.

For mileage programmes that lack substantial emotional added value, the other way to create a wider gap between the value of the points and their cost is simply to reduce the cost of operating the programme. Smart cards have already proven their capabilities in this area.

Some of the first smart card-based loyalty programmes kept the customer's points or frequent flier miles stored in the card's chip. Cardholders could then know their card balance every time the card was used, rather than having to wait for a statement to be sent by mail. By effectively providing a loyalty account statement each time the card was used, the customer immediately knew that the rewards had indeed been correctly added to the card. So they no longer needed to call customer service to ask why last week's flight didn't appear on the statement just received by mail.

Boots the Chemist, a British pharmacy chain, issued over eight million smart loyalty cards to customers the first year of the scheme's operation. All 1260 stores were equipped with smart card terminals capable of calculating loyalty points based on a purchase amount, updating the points in the chip and printing out a loyalty statement with the updated points total. The card is estimated to have cost Boots £52 million over three years of operation, an investment that was more than offset by a sales uplift of over 4%.

Table 2.1 shows Boots' investment over the first three years of their smart card loyalty programme.[26]

Table 2.1 Boots the Chemist smart card loyalty programme

	1997	*1998*	*1999*
Marketing	£8m	£8m	£8m
Capital investment	£4m		
Database	£4m	£2m	£2m
Other	£3m	£3m	£2m
Cost of cards	£6m	£1m	£1m
Total investment	£25m	£14m	£13m
Total investment as % of sales	0.5	0.28	0.24
Sales uplift required to achieve breakeven (%)	4.5	2.5	2.2
Actual sales uplift observed (%)	4	4	4

In general, rewards programmes like Air Miles or Nectar that don't use real-time techniques are estimated to typically cost up to 0.5–1% of sales – in other words, double or triple Boots' investment – due to the substantial costs of managing large databases, printing and distributing rewards catalogues and processing reward redemptions by phone and mail. Although Boots spent a total of £52 million, a similar programme using non-real-time techniques would have cost between £100 million and £150 million and would probably have resulted in a smaller uplift in sales.

Also in 1997, AOM French Airlines launched a similar smart card programme targeted towards frequent fliers, essentially replacing the company's existing 'Carte Capital' and extending it to new passengers. AOM's management was surprised at the cost reductions. AOM's general manager, Jean-Marc Janaillac, said:

> Thanks to instant statement printing at each transaction, our smart card system allows for improved service to our customers, while at the same time doing away with frequent statement mailings and the related call centre necessary to answer the inevitable questions concerning these statements. For us, the smart card clearly represents major cost benefits when compared to the magnetic strip card.

AOM's ticket and check-in counters were equipped with payment terminals equipped with smart card readers. The functionality is simple by today's standards. When the passenger checks in, the Carte Capital smart card is credited with a number of points based on several parameters such as destination and class, and a receipt is provided to the customer indicating the total number of points accumulated. The terminal prints a detailed statement after each transaction, showing the complete status of the cardholder's Carte Capital account (prior balance, points earned with this transaction, new balance, promotional and incentive messages, and so on). AOM's customers always knew how many points they had. AOM uses the statements as a direct marketing tool. Printed receipts eliminate the need to mail out costly account balance statements, resulting in significant savings on direct mailing expenses. The programme functions on a light and cost-effective IT infrastructure. In addition, promotional campaigns can be launched at any time to offer bonus points – for example, during the month of May 1997, AOM doubled the number of points credited on all domestic flights. The total number of points is stored in the card's microprocessor chip. At any time and at any AOM sales counter, acquired points can be debited in exchange for free tickets or instant upgrades.

These programmes were launched a few years before EMV migration, so they required a relatively high cost up front, during the first year, due to the cost of the cards themselves and the cost of upgrading the payment terminals and processing infrastructure. Overall programme operating costs tended to go down significantly once the programme was launched and the initial infrastructure deployed. Today, using an existing EMV payment infrastructure with full EMV features and co-branded Boots or AOM payment cards, the initial investment would be cut by half. The first-year investment would essentially be limited to marketing costs as opposed to card and terminal infrastructure. This is precisely what banks are experiencing today. Akbank operates a full-featured EMV service for over 80 000 merchants with only two additional people assigned to the service on a part-time basis, and uses off the shelf EMV smart cards and standard terminals which they needed anyway in order to become EMV compliant.

DISTRIBUTED VERSUS CENTRALIZED PROCESSING

When NASA began landing unmanned roving devices on the moon, it quickly discovered that centralized control from Earth did not work well. The one-minute delay between an Earth-based central command station and a robot about to go over a cliff meant that robots had to be autonomous. Intelligence had to be placed directly in the device to allow it to make decisions quickly without detailed guidance from Earth.

It turned out that troops of small, autonomous robots built for specific chores, like searching for minerals or preparing landing sites, were cheaper and more robust than a single, general-purpose robot centrally controlled from Earth. These robots can be built quickly and easily from off-the-shelf parts. They are inexpensive to launch. Once released, they can perform many useful services without the need for constant supervision.

Scores of intelligent devices are already becoming part of our lives as chips become embedded into more and more everyday objects. A future where smart doors may turn off the lights when you step out, smart stereos may be able to dynamically adjust the balance based on where you are in the room, and smart vacuum cleaners may come out when no one is home, randomly clean up for a while then discreetly go back to their place under a bed or in a corner is not too far off. These types of household devices all have one thing in common: they are not linked to a central command system, a sort of house brain that manages every little detail. Instead, each device has its own (limited) intelligence. It does one thing tirelessly. If the smart vacuum cleaner breaks down the dust might start to collect, but the other devices keep on functioning.

Individual devices might send messages periodically to a unit that monitors and reports on potential problems. The vacuum cleaner might send a standard e-mail message once a day, telling how many miles it has swept up since it went into service, how much dust is collected in the bag, as well as information on potential problems like a wheel being clogged up. In this case, the central unit acts as a monitor, not as a controller. It adds a layer of value on top of individual autonomous devices without getting in their way and without creating bottlenecks.

Kevin Kelly, executive editor of *Wired* magazine, says that:

> The surest way to smartness is through massive dumbness. The surest way to advance massive connectionism is to exploit decentralized forces – to link the distributed bottom. How do you build a better bridge? Let the parts talk to one another. How do you improve lettuce farming? Let the soil speak to the farmer's tractors. How do you make aircraft safe? Let the airplanes communicate among themselves and pick their own flight paths. This decentralized approach, known as 'free flight', is a system the FAA is now trying to institute to increase safety and reduce air-traffic bottlenecks at airports.[27]

Moving intelligence out to the periphery is a paradigm that echoes what happened when electrical power first became widely available. At that time, the Sears Roebuck catalogue advertised a general-purpose electric motor for household use, complete with dozens of attachments and optional add-ons. The motor could be reconfigured to act as an egg-beater, an ice-cream maker, a clothes washer, power shears or a drill. At the time, it was difficult to imagine colonies of little motors built into each individual device. That would have been considered overly wasteful and costly.

To many engineers designing complex technology, the logic of giving up central control is upside-down to the generally accepted approach of applying thorough, top-down control to every step of the process. Although distributed control based on a bottom-up design might be counterintuitive, it nevertheless consistently proves to be less expensive, more robust and more effective in many areas of technology as well as in economic and social systems. Electric motor devices is one example; robotics is another. And the Internet is yet another. Huge global computer networks are based on chunks of peer-to-peer networks. Imagine instead a top-down, traditional network on a global scale. Who owns it? The federal government? A union of nations that adhere to the same trade treaty? You have a problem with your e-mail? Call the Department of E-mail Administration in Washington, DC.

Centralized control is a residue of the industrial age. Distributed control and real-time response to external stimuli have already become mandatory features in many industries. Smart cards are now bringing this same trend to the payment card industry.

FASTER AND CHEAPER REAL-TIME CUSTOMER RECOGNITION WITH DISTRIBUTED CONTROL

Centralized marketing programmes creep along, exploring unmapped territories of customer behaviour, under the constant guidance of a central control unit. They routinely encounter numerous obstacles, and can faithfully report that a cliff is appearing on the horizon, is approaching, is at an uncomfortable distance, is here, now. Information is digested long after it is generated. The central control unit can, of course, change directions and adjust the programme in many different ways, but commands take a long time to reach the faraway programme. Several days or weeks after the programme falls over the edge of a crevice, central control is proud at its ability to provide a detailed, second-by-second account of the disaster. Moreover, central control will continue to check for vital signs, as the programme lies in pieces at the bottom of the cliff, reporting that, yes, the programme is still lying there and that it hasn't moved much over the past weeks and months. The few times something did move down there might prove that the programme is really still somewhat alive. On the other hand, it might well just have been a few good gusts of wind. Nobody really knows.

Centralized customer relationship programmes are usually unable to react in real time to a customer's behaviour. New commands are sent out via direct mail messages that generally arrive after it is too late. Attempts to speed up communications have resulted in tremendously complex and expensive systems that are notorious for their tendency to collapse under their own weight.

Monthly statement mailings, rewards catalogues, data processing staff, database experts operating sophisticated segmentation software, customer service centres dedicated to statement inquiries, warehouses filled with rewards inventory, rewards fulfilment staff ... that's the expensive marketing infrastructure that is required for a typical centrally commanded programme to function properly. To cut costs, a common response is to send out fewer statement mailings, or print and distribute fewer catalogues. This effectively reduces annual costs, but it only makes the problem worse, since it increases the already unacceptable delay between when something happens in the field and when the central marketing structure finally responds to it. Another way to reduce expenses is to negotiate prices with suppliers to lower the cost of rewards. But a bank is not a retailer. Trying to act like one is inefficient. Often the result is lower-quality rewards.

Another reason why card marketing programmes have become more difficult to design from the top down is that more and more banks depend on merchant partnerships to increase the value of rewards provided to cardholders. In the past, participating merchants could simply offer double or triple points for purchases in their stores. This has changed. Merchants don't necessarily want to offer the same rewards across all categories of purchases, and issuers increasingly need to differentiate against their competitors who virtually all already offer such programmes. A single programme with the same rules across all participating merchant partners doesn't work as easily anymore. Each individual merchant's needs are different. This is, of course, true for large, multi-participant programmes based on mileage, where customers receive one mile for every dollar spent. It is also true across a single chain, since each store in the chain has its own distinct set of problems, competitive pressures and opportunities. Think of a franchise environment where each outlet belonging to the chain is owned and managed by a different person. The individual store manager is usually the person best suited to responding quickly to local market requirements.

Robustness and order emerge from systems that are designed from the bottom up. In his book, *Complexity: The Emerging Science at the Edge of Order and Chaos*,[28] Mitchell Waldrop writes:

> Since it's effectively impossible to cover every conceivable situation, top-down systems are forever running into combinations of events they don't know how to handle. They tend to be touchy and fragile, and they all too often grind to a halt in a dither of indecision.

He suggests that to build robust systems, one should use local control instead of global control. Let the behaviour emerge from the bottom up, instead of being specified from the top down. And while you're at it, focus on ongoing behaviour instead of the final result. Living systems never really settle down. The control of a complex adaptive system tends to be highly dispersed. There is no master neuron in the brain, for example, nor is there any master cell within a developing embryo. If there is to be any coherent behaviour in the system, it has to arise from competition and cooperation among the agents themselves.

Visa itself was designed based on these concepts. Dee Hock, Visa's founder, came to the conclusion that:

> It was beyond the power of reason to design an organization to deal with the complexity of the credit card industry, and beyond the reach of the imagination to perceive all the conditions it would encounter – yet, evolution routinely produced much more complex organisms with seeming ease.[29]

It became apparent that the organization Dee Hock was creating (later to become known as Visa) would have to be based on biological concepts and methods very different from the traditional hierarchical command-and-control type organizations that Hock says 'were not only archaic and increasingly irrelevant, they were becoming a public menace, antithetical to the human spirit'. Hock's organization would have to evolve, in effect, to invent and organize itself. Visa was painstakingly designed based on Dee Hock's belief that 'simple, clear purpose and principles give rise to complex, intelligent behaviour – complex rules and regulations give rise to simple, stupid behaviour'. Visa and MasterCard are highly decentralized, with authority, initiative and decision-making all pushed out to the periphery of each organization, to the member banks themselves.

Likewise, robust customer marketing programmes are more powerful and easier to operate when they are designed from the bottom up. Intelligence must be moved away from the centre and placed where it is most useful, directly at the point of purchase, so the marketing programme can become autonomous and respond to customer behaviour in real time. Real-time technology provides customers with higher-quality service and immediate access to benefits, without all the usual administrative overhead. Programmes capable of real-time customer recognition cost less and provide better service than programmes that are not capable of real-time recognition. Smart cards are the key to achieving this on a massive scale.

REAL-TIME RFM

RFM is recency – frequency – monetary: a method of recognizing different spending patterns based on how recently a customer has used their card in a specific manner, for example at a store, chain, ATM network and so on, how frequently the card has been used in this manner over a defined period of time and how much monetary value has been spent in that time.

One of the most popular customer-centric features offered by EMV smart cards is the offline delivery of targeted advantages such as a free gift or discount after a specific number of visits or dollars spent by the customer in the same month, or varying the rate of points attribution based on cumulative purchases, for example by offering double points when monthly purchases at Carrefour exceed $250. This uses EMV's ability to store recency, frequency and monetary value parameters (RFM parameters) for each merchant's programme within the chip.

Although the RFM customer segmentation method is well known by database marketing professionals, transferring the process to smart cards proved to be difficult. Database professionals traditionally use RFM to rank customers based on the recency of their last visit, the frequency of their visits in a given period and the cumulative monetary value of their purchases. For example, customers might be assigned a recency score of 24 if they have shopped within the last 4 weeks, a score of 12 if they shopped between 4 and 6 weeks ago, 6 if they shopped within the last 9 weeks and a score of 3 if they shopped within the last 12 weeks. Similarly, scores will be attributed based on the total amount the customer has spent (for example, a monetary score equal to 10% of the total amount spent) and the number of visits made (say, four points for every visit made in the year). The method results in a sorted list of customers based on their overall score.

Once customers are compared to each other and ranked in this fashion, a marketing strategy can be developed that takes into account the behaviour of best customers versus worst customers. Offers are formulated differently for each group and are sent by mail to each customer.

The known RFM method requires access to a full customer database with detailed transaction data covering each customer's purchase history. It was designed to function on a central computer system that processes data that includes at the minimum the date and amount of each transaction linked to a specific identifier that uniquely identifies each customer. The customer's identity is usually known via an ID placed in the customer's magnetic strip card that is swiped at the checkout register, allowing the customer's ID to be linked to the purchase transaction. The transaction can be uploaded instantly (in an online configuration) or nightly (in a batch mode).

To a person skilled in database marketing, smart cards do not inherently provide an advantage over other methods that allow a customer's identity to be linked to a purchase. Magnetic strip cards work fine. So do simple paper cards and keyrings with the customer's ID printed in the form of a barcode. Supermarkets have been using barcoded customer cards and keyrings for almost twenty years. The card is scanned at the point of sale terminal, just like an item's Universal Product Code (UPC). The customer's ID becomes part of the purchase transaction and is uploaded to the store's server along with the products that were purchased by the customer. There, a marketing database management system can classify and segment customers based on their purchase behaviour. Average purchases on a per household basis can be calculated and marketing plans formulated to encourage specific categories of customers to change their purchase habits. Letters are sent out to selected customers, for example, offering them a free turkey if they spend a minimum of $500 before Thanksgiving. This approach works quite well when the store manager can get into the complexities of database marketing. Smart cards are not required, so database marketing specialists have rarely attempted to use smart cards to store and process RFM behaviour parameters.

Although the common approach described above, using barcoded cards, does not rely on real-time methods, many PC-based point of sale scanning terminals have nevertheless offered real-time features for many years. A feature commonly offered since the early 1980s is the ability to use a customer's barcoded card to trigger the terminal to give specific discounts reserved for that customer's specific category. Customers who sign up for the merchant's loyalty programme receive special discounts and a monthly newsletter. Some of the most effective features available with modern point of sale scanning terminals don't even require a customer to present a loyalty card. After scanning all the purchases in the customer's shopping basket, the point of sale terminal goes back and adjusts prices if the overall transaction is over $20. This is how systems handle the dual pricing mechanism used in many US supermarkets, offering convenience store pricing to those customers who just buy a couple of items, and discount store pricing to those who buy more.

Some systems are even capable of storing the customer's loyalty points on the same in-store server that provides price look-up information every time an item is scanned at the point of sale terminal. This effectively allows some customer information to be available in real time at the point of purchase.

There are a few drawbacks. For one thing, real-time access to a customer's behaviour across multiple stores belonging to the same chain is often difficult and costly. Also, the type of marketing analysis that segments customers based on their purchase history would quickly bog down even the largest and most powerful servers available today, if analysis is attempted in real time. Centralized systems work fine when rewards are sent out by mail, but they prove to be inadequate when rewards need to be delivered in real time. A different approach is required, more robust and more easily capable of performing complex marketing analysis in real time. In the late 1980s, smart card pioneers were already promising that the new technology would one day allow this.

The traditional RFM method could not be transferred to a smart card without major modification to the method in its implementation and even in its spirit. It was necessary to recombine smart card technology with the RFM method in a completely new, innovative fashion in order to arrive at a useful result. The difficulty of using smart cards as a loyalty marketing device is illustrated by the number of companies that have made such attempts but for various reasons fell short of fully using the card's real-time capabilities.

In 1988 a US company, Advanced Promotion Technologies, developed an interactive electronic marketing system delivering targeted promotions, a frequent shopper programme, financial services and other information to shoppers in supermarket checkout lanes. Terminals used full motion video and stereo sound to present the consumer with informational messages while touch-screen displays were used for consumer interaction. The system initially planned to use smart cards for the frequent shopper programme. The company was well positioned to leverage smart card technology and create innovative methods to build customer loyalty. Shareholders included packaged goods manufacturer Procter & Gamble and a well known database marketing firm. And yet, in spite of all this, the company never did succeed in using the smart card as a repository device containing dynamically processed customer behaviour information. One reason was that they were trying to store all of the customer's prior purchase data within the chip, rather than storing more manageable, condensed data snapshots. Smart card technology was finally abandoned as the company moved to a completely online architecture requiring satellite communications. The project never achieved the ease of use and cost-efficiency necessary for a massively deployable mainstream solution.

HTEC, a UK firm, is a leading supplier of loyalty terminals to petroleum companies. Using magnetic strip read–write devices at the point of sale, HTEC's system is capable of dynamically storing information directly on the card's magnetic strip. Reading information from a card's magnetic strip is easy; you simply have to swipe the card. Writing information is much harder. You have to swip once to update the magnetic strip and then swipe a second time to make sure the update succeeded. If an error occurred, the card would need to be swiped another two times. This is why data is written to magnetic strip cards using motorized devices that can automatically swipe the card multiple times without operator intervention. But motorized card read–write devices are expensive and prone to physical problems like jamming. The overall magnetic strip card system's economics are therefore based on expensive, motorized read–write terminals and low cost magnetic strip cards. The business case makes sense in situations where there are few terminals but many cards. Although the customer's total acquired loyalty points are stored in the magnetic strip, other information related to the customer's behaviour, like cumulative spend in a given period, or frequency of visits, must be stored on an external database server. Storing sensitive data on magnetic strip cards poses another significant problem: the cards are relatively easy to clone. Get two cards, load points up on one gradually, then when there are lots of points saved up, copy the card's magnetic strip contents to the other card, redeem the points on the first card, go back home and recopy the full card onto the now empty one, and start over again. You don't have to understand each byte of what is stored in the magnetic strip; you simply need to know how to clone the full contents.

With the advent of smart cards, which offer greater security, as they are much more difficult to clone, it became apparent that a number of problems could be solved at the same time. If the customer's detailed behaviour parameters could be placed in the smart card's microprocessor chip and dynamically processed in real time at the moment of purchase, it would become much easier to offer customer rewards based on actual purchase behaviour.

Easier said than done. In 1995, my company, Welcome Real-time, was asked by France's Ministry of Youth and Sports to help convert the Carte Jeunes youth card to a chip card, so as to offer young people in France more benefits than what was previously possible, when Carte Jeunes was a 'flash card', that is to say a card which was simply shown to a participating merchant to obtain a discount. At that time, by 1995, Shell in the UK and Takashimaya in Singapore had introduced scheme-wide currency programmes based on chip cards which removed the need to send statements or redeem awards by mail. These deployments were well known at the time and were considered as potential models for Carte Jeunes.

Our work with Carte Jeunes progressed through a number of stages. First there was a system based on scheme-wide Carte Jeunes currency in which the customers' points counter resided in the card's chip as well as on a central database. All participating merchants would give points in the form of Carte Jeunes currency which was exchangeable for payment at any participating merchant. The balance of the currency was contained in the card and could be provided at each transaction, thus eliminating the need for statements or reward request forms. However, France's central bank, the Banque de France, considered that the proposed Carte Jeunes currency was too much like real money and could compete with cash in the French economy, so it would need to be regulated just like cash. This was a daunting challenge which we decided we would prefer to live without. Apart from that problem, investigations revealed it was going to be difficult to persuade many small merchants to participate in universal point schemes. Some felt that they were being asked to provide a deferred discount which would probably be used to purchase something in another merchant's store. A number of merchants preferred to stay with the flash card concept, offering discounts immediately on purchases in their stores.

A second stage was a proposal that points acquired everywhere could be redeemed only at participating cinemas. Merchants found this more attractive, as the points then had an emotional element added to their value – they were promoting cinema and culture. Perhaps the Ministry of Culture could even increase the value of the points. This proposal was not implemented because Carte Jeunes felt that a currency redeemable only at cinemas was of limited benefit to cardholders. It wanted to provide cardholders with a much larger selection of redemption options.

These examples illustrate the difficulty in developing a system which linked together many diverse merchants with different technical infrastructures and promotional methods. Since a universal scheme-wide currency proved impractical, attention turned to exploring systems operated by single chains where points accumulated within the chain did not mingle with points accumulated elsewhere and where specific rewards could be provided for specific merchants or groups of merchants. We became aware that to implement multiple distinct merchant programmes on a single chip card, the chip card-based system needed to be capable of using multiple counters within the chip, each counter dedicated to an individual merchant or chain of merchants.

The world's leading supplier of chip card technology was the French company Gemplus. Gemplus proposed a Carte Jeunes chip card with several points counters. However, the problem remained that, with multiple loyalty programmes, if a chip card was intended to function with loyalty programmes A and B, the card would have to be produced with loyalty programmes A and B predefined within the chip's data memory. Each merchant participant would have to ensure that they used only the memory space which had been assigned to them. This allocation could not later be amended. A customer's chip card would carry all available loyalty programmes regardless of whether or not the customer participated in all of them. With the then available technology there was a limit of about ten programmes which could be loaded to pre-assigned memory locations for merchants. We recognized that such types of chip card would not be able to provide a system which could be implemented in an environment consisting of potentially thousands of different merchants running their own distinct loyalty programmes. Gemplus had no other solution to offer.

The next proposal made by Gemplus to Carte Jeunes was to create co-branded chip cards for local merchants. For example, a chain of cinemas in Paris would issue their own Carte Jeunes cards with the name of the cinema on the cards as well and their dedicated loyalty programme memory zoned within the chip. However, that would effectively limit a merchant's loyalty programme only to those cards which were co-branded with the merchant. Most merchants wanted to provide loyalty rewards to all Carte Jeunes cardholders in their geographic area, a much larger set of cardholders than any co-branded programme could possibly attract.

Our thinking at the time as to these practical limitations was that the POS terminal of a particular merchant would have to be configured to read and write to a single memory zone or file with a specific address with which it had been predefined. For example, merchant A's terminal would be programmed to always function with memory zone A in the chip. Thus the number of merchants that participate in the loyalty system was limited.

We had a critical 'eureka' moment when we realized that our thought process, and that of the rest of the industry, was in fact upside down and needed to be looked at in a completely different way. The smart card industry only saw value in information present in the chip. We suddenly understood that the solution to our problem came down to seeing that absence of information is also important.

The light bulb flashed when we were trying to look at implementing a system where customers can go to a large number of restaurants, for example, and each restaurant wants to give a free cocktail to customers on their first visit. Then on subsequent visits some restaurants might decide not to offer anything else while others might then offer a free dessert after the customer's third or fourth visit. The thinking at the time was that the card issuer would need to know beforehand all of the restaurants that they would be working with and have them all in the chip so that the POS terminal can look into the chip to see if the customer has been to a particular restaurant before or not. No normal smart card at the time was able to store the tens of thousands of potential restaurants each customer might go to. But in fact, we realized that if there is no information in the card relating to a specific merchant, that signifies a lot. If a customer goes to the restaurant and the POS terminal scans through the behaviour file in the chip, and sees that the customer has not been there before, because there is no information concerning this restaurant stored in the chip, then that is significant information because it's saying that it's the customer's first visit there. So it's the customer's first visit. This triggers the free cocktail. The information is added into one of the slots and then the merchant can either let it stand there and no longer give anything else, or if the merchant's programme is then to give a free dessert at the customer's third or fourth visit then points information is added to count the number of visits and then trigger the free dessert. It relates to the behaviour even if it is empty, because the POS can know that the customer has not been to this restaurant. It's the lack of information that suddenly appeared very important and useful. Prior to that the information had to be loaded in the card, the customer had to have the coupon for that cocktail to be able to pull it out when he goes to the restaurant. That's how most people were thinking about loyalty programmes as well. If a customer wants the loyalty programme for the restaurant, somehow it needs to be in the customer's card before it can be used.

In January 1996, we eventually used this realization to engineer a breakthrough patent using smart cards to create a distributed database, placing behaviour information directly in the customer's wallet and direct marketing software in the merchant's terminal:

- Behaviour data, including RFM parameters, was calculated, stored and dynamically updated in real time on a smart card, as opposed to being calculated, stored and updated centrally in batch mode. The merchant's programme was added to the behaviour file by the POS terminal the first time the cardholder used the card at that merchant's store. Behaviour information and a points counter for merchants that the cardholder never visits are never added to the chip card. There was no pre-allocated memory slot for such non-visited merchants. Any chip card could still have just a few points counters but the first points counter of any one card would be allocated to one merchant, while the same points counter on another card could be allocated to another merchant. The approach of the invention was to allow chip cards with a small memory capacity to be used across thousands of merchants each offering their own distinct incentives.
- The ranking process at the heart of the RFM method was dispensed with since, in an offline mode, real-time access to a full customer database is not possible. Each customer transaction must be analysed independently of all other customers, which is contrary to the traditional RFM process.
- The ranking process was replaced by an algorithmic processing at the POS terminal that reacts in function of each customer's behaviour profile in order to determine whether or not to print a coupon.
- A coupon containing a message or an offer was automatically printed out directly at the POS terminal based on behaviour data in the card.

This was originally conceived as a method of enhancing mileage points programmes by adding the ability to vary the number of points issued based on RFM behaviour parameters, as opposed to simply adding points to the card as a fixed percentage of each purchase. Upon marketing the system, it was discovered that the process was applicable to another, much wider use: the replacement of paper loyalty cards that are punched, stamped or otherwise marked at each visit, providing the customer with a free item or service once the card is full.

A single file in the smart card's memory can hold numerous RFM slots, each dedicated to a merchant's electronic incentive programme. The process can be made very efficient; 1000 bytes of memory in the smart card's chip can store up to 60 distinct RFM combinations, for up to 60 distinct merchant incentive programmes. This very small footprint can be supported by many low-cost smart cards which were originally intended as single-function credit or debit cards, and which effectively become very low-cost full-featured enhanced payment cards.

Allocating a slot to a particular merchant can be done automatically when the card is first presented at that merchant's terminal. The merchant's RFM slot will remain active as long as the incentive programme has not expired. After expiration, the slot becomes available once again. The customer 'chooses' which merchant slots are active by shopping regularly at those stores, and 'deactivates' slots by not shopping there anymore. The operation is completely automatic, for greater speed and minimum confusion. It is always best not to encumber the flow of the transaction at the point of purchase through asking customers if they want to join the programme. Merchants should only be required to enter the amount of the transaction, just as they already do, leaving the software to perform complex marketing analysis on the customer's behaviour data in real time, without requiring additional keystrokes or the use of complicated function keys.

In 2001, the method described here was at the heart of a landmark judgment of the Federal Court of Australia, the first so-called business method case to come before an Australian court. The court held that the invention was rightfully patentable and that it had been infringed by competitor Catuity Inc. The judge found the patent to be inventive and not obvious. In relation to obviousness, the judge stated that:

> One cannot help but be influenced by a consideration of the field of the invention. This field concerns the application of state of the art information technology to large scale retail operations in leading industrialised countries. The financial stakes were high. We are not concerned with an improved model of button up boot. The vast array of prior art shows that many people – no doubt highly talented – were working in the field without coming up with the invention. It is hard to ignore the comment that if what the inventors did was all that obvious, someone else would have done it.
>
> *Welcome Real-Time SA* v *Catuity Inc.* [2001] FCA 445, 17 May 2001

MEASURING TRANSACTION RICHNESS

Traditional methods using barcoded cards, social security numbers, or magnetic strip payment cards are limited in their ability to address customers differently at the point of sale terminal. Their primary function is to collect customer data and upload it to a host. Any additional demands made on the system are difficult to process.

As cards and terminals become more intelligent, payment transactions become far more information-rich and can now respond differently to each customer, based not only on data concerning the customer's current transaction but also on data concerning many prior transactions. All of this information is available in the card, so the terminal can perform many more functions without slowing down the payment process.

The merchant's credit card point of sale terminal can now deliver not only coupons for free items or discounts, but also targeted information similar to direct marketing letters, customized according to each customer's actual behaviour. All of these features can be used independently of each other, or in unique combinations defined by the merchant. This all happens at the point of sale terminal, in real time.

Richer real-time transactions become a tool that helps store personnel better welcome their customers. In general, merchants and customers will both tend to prefer the payment method with the highest level of transaction richness.

Payment systems will differ widely in the level of richness they bring to each transaction. Magnetic strip systems have a low level of transaction richness. This is not to say that a central database management system cannot add richness to uploaded transactions. This indeed happens with data mining software. What we're talking about, however, is real-time richness, at the point of sale terminal, as the merchant is speaking with his or her customer. Smart card systems offer a potentially much higher level of transaction richness. Although this is intuitively obvious, it would be useful to be able to quantify the richness provided by each of the different payment methods.

The Transaction Richness Quotient (TRQ) is a measurement tool that can rigorously quantify the added value a particular payment method provides in real time at the moment of purchase. TRQ is calculated by analysing the various parameters that are processed during the payment transaction. Parameters include information generated by the terminal – such as the purchase amount, the date and time of the transaction, as well as information provided by the customer's card – for example the cardholder ID, the date of this customer's prior transaction in this store, the number of transactions made during a specific period and the cumulative amount spent.

The objective of real-time marketing at the moment of purchase is to increase sales and profitability. This is accomplished through three complementary objectives:

- Customer knowledge generation – the ability to generate useful knowledge about the customer and have it instantly available at the payment terminal.
- Reward attribution – the ability to attribute rewards to the customer immediately, such as discounts or free items.
- Relationship building – the ability to establish a long-term relationship that is not necessarily based on financial rewards but rather on 'soft' benefits like preferential treatment.

These three objectives should be developed concurrently in order to have a well-balanced system. An unbalanced system may not have the desired impact on the customer's behaviour. For example, a system can go overboard on knowledge and neglect reward or relationship. This typically happens with data mining systems that analyse payment transaction data. The system itself generates knowledge, but all alone it is unable to have significant impact on how much the customer spends. The results of data mining systems must always feed other activities like database marketing in order to attribute rewards (by sending them to the customer) or build a long-term relationship. TRQ seeks to measure a particular system's ability to meet all three of these basic objectives concurrently, in real time, at the moment of purchase.

Each parameter that is processed at the payment transaction level will impact one or more of these objectives. A parameter that only impacts 'Knowledge generation' will receive a score of one. Another parameter that impacts 'Knowledge generation' and 'Reward attribution', for example, will receive a score of two. Parameters that simultaneously impact 'Knowledge', 'Reward' and 'Relationship' have a score of three. After analysing all parameters, the scores are added up. This is the overall TRQ.

Individual card implementations can be compared to one another in terms of each system's TRQ. Subtleties like a particular system's capability in building relationships, as opposed to attributing financial rewards for example, become evident when one looks at the weight of each set of parameters within the overall TRQ.

A large majority of loyalty programmes are copies of airline frequent flier programmes: earn a mile or a point for every dollar spent. Present your card at the checkout where your customer ID gets linked to the transaction. You know how much you spent today and perhaps the receipt will tell you how many new points today's purchase entitles you to. That's about it. You might have accumulated a large number of points, or you might be in the store for no other reason than the store you usually shop at is out of a particular product. The merchant has no way of knowing. Systems with a low TRQ are like navigating through clouds without a guidance system and without the ability to react in real time to avoid flying into a mountainside. Traditional systems let you know lots of things about the customer once the data has been centralized and processed, but by the time the central system has come around to processing the data concerning a customer's specific visit, the customer has already left the store, gone home, worked a few weeks, and shopped over and over again since the date of the specific transaction that the central system is currently processing. The TRQ is a way of measuring a system's ability to react in real time to a customer's actual behaviour.

The first smart card-based loyalty programmes launched by large retailers like Shell and Boots in the UK, or by AOM French Airlines in France, were able to go a step further and increase the TRQ by keeping the mileage account on the customer's smart card. This immediately provided improved visibility. The customer could now know, at each transaction, exactly how many miles had been accumulated. The miles are instantly available, so the day a Boots customer finally has enough to buy an expensive perfume or lotion, they don't need to call up Boots, write a letter, wait for a gift voucher and so on. They simply present their card along with their purchase and the points are debited from the chip.

The customer's perception of the quality of service provided by these new programmes has soared. It's easy to know how many miles you have since, at every visit to Boots, your receipt tells you where you are, and because you know how many points you have, you don't have to call up a phone operator and ask for your balance. Unlike traditional methods, a smart loyalty card does not increase the number of customer service people required to answer account-related questions. Another major benefit is that points can be redeemed by presenting the card at any point of sale and having the points debited from the chip. The customer doesn't have to request a gift voucher, and the retailer doesn't have to put in place a complicated paper voucher management system.

IMPROVING LOYALTY PROGRAMMES BY INCREASING THEIR TRQ

The Boots, Shell and AOM programmes are significantly better than programmes with little or no real-time visibility, but they still don't give a clear enough view from a marketing standpoint. They still don't take into account a customer's actual behaviour. Proportionally speaking, all customers, whether they shop frequently or not, earn the same number of points. How can the TRQ be increased with these programmes? If Boots were to process recency, frequency and monetary value parameters at the chip level, customers could be rewarded based on tiers. The first £50 spent in a given period might earn 2% in points, the next £50 might earn 3%, while anything over £100 might earn 4%. This would result in a significant increase in transaction richness. There is an additional benefit: cost savings. By giving less to casual customers and more to frequent customers, the overall programme is less costly than giving everyone the same 3%.

Multi-merchant programmes allow customers to earn points at a number of retail chains. Again, as in the Boots programme, points are given based on a simple purchase amount. Using the electronic promotional incentive concept, an individual merchant participating in the programme can vary the points based on usage of the card at that particular merchant, for example by offering 2% in points to anyone who walks in the door, and by offering an additional 3% to customers who spend a minimum of £100 in a single month.

Many franchise chains will find that electronic incentives allow them to create a chain-wide programme that finally satisfies individual store owners. With many franchises, the store owner feels he's building loyalty to the chain, as opposed to his particular store. If the operating costs and the rewards are completely paid by central marketing, that's alright. The budget is coming out of whatever each franchisee pays in marketing fees to the franchiser. When a franchise chain's central marketing tries to create a programme in which rewards are financed by individual stores, the project usually falls apart. 'Why should I give a free burger', asks a restaurant manager, 'to someone who earned his loyalty points somewhere else? I want to give a burger to customers who earned their points in MY restaurant.'

This is a common complaint. Merchants that participate in multi-store or multi-chain programmes want to give loyalty points to frequent customers, but they want those points to be redeemed in their store, not some other store.

Electronic incentives solve this problem, as each store can offer its own electronic incentives. As a customer, you receive your card initially empty. When you go to a store that accepts the card, that store's electronic incentive programme is automatically added to the chip. The receipt tells you that after three more visits to this restaurant, for example, you will receive a free burger. When you go to other participating stores, their incentive programmes will also be added. This allows each merchant to trigger incentives based on the customer's prior purchase behaviour at that merchant.

Financial institutions involved in private and co-branded card programmes know that the market for traditional 'frequent flier mileage' type programmes is saturated. Today, smart cards offering real-time mileage points accumulation, point of sale redemption of points and instant delivery of electronic incentives, can open new markets for numerous franchise chains that could not adapt traditional catalogue-based mileage programmes to function in their environment.

MEASURING THE RICHNESS OF PAYMENT TRANSACTIONS

Cash and cheque transactions can also be measured. The TRQ for cash appears intuitively to be close to zero, in any case well below cheques and e-purse. Cheques can be used to generate knowledge about the customer; for example merchants occasionally collect their customers' names and addresses from cheques. But with cash, a merchant doesn't know who the customer is. So the TRQ for cheques would appear well above that of cash. In reality, it's just the opposite in some cases. How can cash sometimes have a higher TRQ than cheques? Cheques can generate knowledge, but cash can sometimes generate rewards to the customer. Although it is illegal, electricians, plumbers and gardeners in the south of France can routinely quote you two prices: the normal price and the cash price which avoids VAT, income and professional taxes.

Many e-purse pilots have been launched to compete with cash and cheques. Most have failed. The primary reasons have been that customers and merchants find e-purse to be too much of a bother. Another way of looking at why they failed is by measuring the richness of e-purse transactions. What is the TRQ of an e-purse? It is around the same level as that of cash, unless cash provides a discount reward that e-purse does not, in which case cash may in fact have a higher TRQ than the e-purse.

Customers will tend to prefer the payment method with the highest TRQ. If you are launching a debit card or an e-purse in a specific environment where cash has a particularly high TRQ, you must take extra care to make your card product attractive.

Magnetic strip credit card systems generally have a TRQ of around two. Electronic cash implementations like Visa Cash, Mondex or Proton have a quotient of around five, comparable to France's chip-based debit card system. An early implementation of Welcome Real-time's software for smart cards had a TRQ of 23. By the end of 2000, the latest version of the software had a TRQ close to 100 and by 2004, the latest version's TRQ was at 270.

Fifteen years ago, magnetic strip payment cards were everywhere. The financial industry then began moving to smart cards, initially in France for debit cards, then in other parts of Europe. Card issuers soon began using smart cards to implement loyalty point mileage systems. More recently, issuers have begun using smart cards that let merchants deliver their own targeted incentives based on the cardholder's prior shopping history.

The TRQ has been increasing rapidly over the past years. Meanwhile, transaction times have been cut dramatically; from one or two minutes with magnetic strip credit cards (for an online authorization and manual ID and signature verification), transactions have come down to five to ten seconds with smart cards. Something quite similar to Moore's law (the observation by Intel founder Gordon Moore that the power of microprocessors is doubling every 18 months) seems to be applicable here. Once you think about it, there's no reason why the payment industry should not be governed by principles similar to those that drive the microprocessor industry. In light of the huge potential for new, complex features that smart card software will provide in the future, the TRQ will continue to increase over the next few years.

THE IMPACT ON CARD AND TERMINAL HARDWARE

The TRQ is a standard way of measuring the richness of a particular payment method. It does not measure how the payment transaction physically happens, such as how long it takes or how much investment is required in infrastructure costs.

From most merchant's point of view, cash is fast and requires no investment in infrastructure. The cost of handling cash is an ongoing expense that most merchants don't quantify. Credit cards processed online for authorizations are slower and require an expensive infrastructure.

Individual card technologies, terminals and service providers can have an impact on the speed of transactions and their cost. Calculating TRQ ratios per second and per dollar for each individual combination of infrastructure elements can help in choosing the right suppliers. A specific terminal model that offers fast transaction processing and printing will have a higher TRQ per second ratio than other terminals. This is an important ratio because point of sale payment transactions, no matter how rich, must be carried out in less than a handful of seconds.

CREATING COMPETITIVE ADVANTAGE THROUGH A RICHER PAYMENT EXPERIENCE – FIVE KEY OBSERVATIONS

1. The moment of purchase is one of the very few moments that a customer is holding your card in their hands, touching it, getting ready to use it. The rest of the time it's in the customer's wallet and probably out of their mind. A remarkable opportunity exists to provide an extraordinary customer experience at the moment of purchase, and become truly differentiated in the eyes of your customer. The experience a customer enjoys while using your card to pay will become the next competitive battleground.

2. Most merchants offer paper incentives to their customers, such as coupons distributed in the Sunday paper or printed in leaflets stuffed into letter-boxes. Rather than distribute these incentives blindly, merchants can use your EMV infrastructure to provide their highest-value promotions to their best customers. EMV cards that keep track of recent customer behaviour can also help to streamline the way soft benefits are managed – benefits that are not necessarily promotional discounts such as a welcome gift, VIP access, special upgrades or advance notice of sales.

3. Virtually all credit cards, and many debit cards, offer mileage points. Today, these programmes are all essentially the same. If the card issuer is capable of creating a wide gap between the value of the points and their cost, the

programme makes sense and has a chance of success. One way to widen the gap is to increase the value of the points. Another way is to reduce the cost of operating the programme.

4. Each merchant's needs are different. Each store has distinct problems and opportunities. The individual store manager is usually the person best suited to responding quickly to local market requirements. Intelligence should be moved away from the centre and placed where it is most useful, directly at the point of purchase. Programmes capable of real-time recognition cost less and provide better service. EMV is the key to achieving this on a massive scale.

5. Merchants and customers tend to prefer the payment method with the highest level of transaction richness. Richness is measured by the Transaction Richness Quotient (TRQ), which seeks to measure a particular payment method's ability to concurrently generate knowledge, attribute rewards and build relationships with customers in real time, at the moment of purchase.

CHAPTER 3

Creating Competitive Advantage through Better Branding

Now that I'm ahead, how can I make it impossible for competitors to catch up?

Senior executive, Latin American bank

The little rectangles of plastic that we use to pay for things would all be identical if it weren't for each card's distinctive brand along with the services that the brand delivers and the promise that the brand represents. When customers decide which cards earn the privilege of residing in their wallets, they don't do it by comparing quality and thickness of plastics, nor the size of the chip or the microprocessor's clock speed. They decide based on the services and features offered by a particular brand. When the brand represents services and features that the customer finds useful, the card has a decent chance of finding a place in the customer's wallet. Once the card is in the customer's wallet, half the battle is won. To succeed, the card must be made irresistible. If the brand is irresistible, the card receives preferential treatment by both customers and merchants. The goal of course is to do everything possible so that customers and merchants feel that they 'must have that card'.

When a bank issues a private card on behalf of a merchant, the only brand that appears is the merchant's. The card can usually be used only at that merchant's chain, although in some cases the card may also be honoured at other stores that have entered into a partnership with the primary merchant, usually when they belong to the same conglomerate. Private cards really work best when the merchant has such a compelling market presence and product offering that customers feel they simply can't do without the card. For the vast majority of merchants, this is hard to do. Private cards tend to be of interest to a small number of already loyal customers.

With a co-branded card, three brands appear on the card, the merchant's brand (for example American Airlines), a payment brand (such as MasterCard) and the name of the bank issuing the card on behalf of the merchant (continuing with this example, Citibank). The issuing bank's brand sometimes appears on the front of the card, sometimes on the back. A new trend is to place the payment brand on the back of the card, as Citibank has recently done in the US. Co-branded cards generally attract a larger share of a merchant's customers than private cards. Because co-branded cards can be used anywhere the card's payment brand is accepted, they are easier to launch than private cards and are more easily adopted by customers. In fact, many private card programmes have recently migrated to co-branded cards.

Branding is a simple way to quickly tell customers and merchants how the card works. The various brands placed on the card help to clarify a potentially confusing product. In the Citibank/MasterCard/American Airlines example, customers understand that they can use the card for payment everywhere MasterCard is accepted. They also understand that they get special rewards with American Airlines. If they want to increase their credit limit or have questions concerning their account, they know they need to call Citibank.

EMV CREATES NEW BRANDING OPPORTUNITIES

The market for private cards and co-branded cards is already mature and branding issues have long been resolved. How about the completely new market of EMV chip cards? EMV offers a richer payment experience with many new payment features at the moment of purchase which were unavailable with prior generations of card products. What impact would such features have on branding? For example, since EMV cards allow thousands of merchants the ability to adapt receipt messages or promotions using behaviour data stored in the chip, how might the card's branding be affected? It would be a mistake to proceed under the assumption that EMV requires no specific thought or effort related to branding or marketing. When a bank limits its EMV implementation efforts to purely technical activities, and leaves marketing and branding out of the process, the bank risks spending a large amount of money just to become EMV-compliant, and could see competitors deploy much more powerful card products. Once a basic, limited EMV infrastructure is deployed, it is harder to go back again later and build in additional EMV features which were left out in the initial deployment. However, building a full-featured EMV infrastructure from the beginning costs virtually the same as a limited, basic infrastructure.

Just as prior card products required specific branding strategies adapted to specific markets, the new EMV market will require its own distinctive branding strategy. Early EMV deployments did not specifically address branding: many used branding strategies inherited from the past. More recently, card issuers have adopted strategies better suited to next generation EMV payment cards. In any case, when a bank migrating to EMV does not take the effort to think about branding, the bank's card will be positioned by others, not by the bank. It is better to control the positioning than be controlled by others.

Merchants that currently offer paper punch cards or paper promotions and that don't currently offer private or co-branded payment cards will not be offering such cards anytime soon. If issuing a private or co-branded card in their merchant category made sense, these merchants would have done so long ago. But linking their existing promotions to your EMV payment cards is another matter altogether. Inviting them to use your EMV infrastructure to target their best promotions to their best customers, determined on the basis of behaviour data in the chip, is an exciting opportunity for many merchants who currently have little or no ability to target their offers. Ideally, they should be able to address the largest possible customer base.

When a merchant runs an incentive programme triggered by an EMV payment card, alongside potentially dozens of other merchants, no single merchant brand will appear on the card. How do cardholders know which merchants offer surprise gifts? How do they know which card to pull out of their wallet? How do they know whether they get a free sandwich at their fourth visit or after spending $20? How do merchants know which cards they honour in their electronic incentive programme? They can't just say 'Get a free sandwich the fourth time you pay with your chip card' because not all chip cards will be enabled for the feature. Merchants might be able to specifically identify a bank, as in 'Get a free sandwich the fourth time you pay with your new ABC Bank smart card'. One drawback to this is that it is cumbersome, especially if ABC Bank has other smart cards that would not qualify – say, ABC employees' cards used to pay at company vending machines and photocopiers. Another drawback is that it potentially limits the merchant's programme to a small portion of customers. It certainly puts the bank in a weaker negotiating position, since ABC Bank is blatantly asking the merchant to promote ABC Bank.

A new brand is better, something like Axess (used by Akbank) or WOW! (used by Mashreqbank). That way, the cardholder and the merchant can refer to the programme quickly. All the merchant needs to say is: 'We give surprise gifts to customers that pay with their Axess card.' The card issuer doesn't have to get into the details of every offer provided by every merchant, and merely needs to say, 'Use your Axess card everywhere it is accepted and get lots of surprise offers.'

A NEW WAY TO PAY

Enhanced payment brands for EMV cards are different from traditional rewards brands which usually evoke a simple counter concept to count miles, points or cash back. EMV brands need to be richer than traditional brands, as they need to evoke a wider variety of customer benefits, not just accumulated cash rewards. Think of traditional rewards brands like Membership Rewards (an American Express brand evoking a car's mileage counter), Air Miles, Visa Buxx or Nectar, with honey dripping slowly, evoking the idea of collecting something sticky over a long period of time. These brands do not modify the way a payment transaction happens at the point of sale, how the payment function is experienced by the cardholder and the merchant. Whether you pay with a normal bank credit card, one with an Air Miles brand on it, or one with Nectar, or even if you use a debit card, the point of sale payment experience is virtually identical. There is little need to place a traditional rewards brand at the merchant's store and indeed few, if any, of such brands appear regularly in stores. But enhanced payment brands for EMV cards are different. These brands indicate a new, richer, enhanced payment experience.

Axess and WOW! are attempts at something different from traditional rewards brands, something closer to payment itself. They do indicate something new and unique happening when you use the card to pay. The payment receipt is different, with more information and sometimes even coupons for free goods or discounts. In many cases, you can even choose to pay with your accumulated points or cash back, directly at the point of sale, instead of charging the sale to your credit card account. These new EMV brands actually appear as acceptance symbols in stores, often next to the traditional payment acceptance symbols, Visa, MasterCard and American Express. They are in fact really enhanced payment brands, as they indicate a richer payment experience than that which customers and merchants are currently accustomed to.

These brands promise to enhance the payment function with fun and exciting features. The Axess brand does not exist without MasterCard, the basic underlying payment function brand; it coexists with the MasterCard brand. It enhances the MasterCard brand on the card, with both brands clearly visible on the front. It adds value to the MasterCard payment function, with new features built into the payment transaction. While the MasterCard brand at the merchant outlet tells customers which cards they can use to pay with, the Axess brand tells customers where they can get special rewards and preferential treatment when they use that card to pay. It transforms the payment *function* into a payment *experience.*

These new brands are attempting to evoke the idea of 'a new way to pay' and tying that to their EMV migration. They see EMV as a new way to pay, not just as a more secure old way to pay. This is a more powerful value proposition and much larger than no brand at all. It is more powerful than limiting the brand to a rewards or loyalty coalition concept. A brand that means a new way to pay in your customer's mind will be easier to sell to merchants. Merchants want to be modern, so if you are successful in creating 'a new way to pay', merchants will feel the need to accept cards with your brand. Think of 'Axess, a new way to pay at Carrefour,' and 'Axess, a new way to pay at Migros,' and 'Axess, a new way to pay at McDonald's'. Today, Akbank's card is accepted at two of the top food retailers in Turkey, Carrefour and Migros, together representing a total of 60% of major retail outlets. A card brand which evokes the idea of a loyalty rewards coalition would probably not have been accepted across the two largest competing food retailers. But 'a new way to pay' is different. An EMV card with no new branding strategy would not have been promoted by both Carrefour and Migros at the same time, and would not have enjoyed the marketing energy and resources paid for by these large retailers. People just don't get excited about an EMV card's basic fraud-prevention capabilities. It is impossible to imagine McDonald's paying for signs and ads in their restaurants inviting customers to 'use your chip card here and enter your PIN code'!

Let's compare two significantly different brand positioning strategies. Both of the strategies we will look at are valid and logical. Both come from the same starting point, but each leads to a very different destination, one limiting and the other expansive. Think of positioning the new card for acceptance at Carrefour. You can position the card brand as 'Carrefour's new loyalty card, Axess'. Or you can position it as 'Axess, a new way to pay at Carrefour'. You already know which one I prefer, but let's go through the deeper implications of each of these.

'Carrefour's new loyalty card, Axess' is essentially asking Carrefour to depend on you for its loyalty card. This is essentially a co-brand or private label discussion. Carrefour is in the driver's seat. Try explaining how you want to allow other retailers to also trigger promotions and rewards from your card (oops, I mean not from your card, from Carrefour's card ...). At best, you might be able to convince Carrefour to be part of a coalition, on condition that you agree never to let Migros join.

In comparison, 'Axess, a new way to pay at Carrefour' presents no conflict with the retailer's own loyalty card, whether it exists or not. Carrefour will see little reason for the card not to be accepted by a wide variety of retailers ('A new way to pay at Migros; A new way to pay at McDonald's; A new way to pay at Wal-Mart ...').

'A new way to pay' positions the card's added value within the payment function – in other words in your world, not the retailer's. It transforms a relatively basic payment process – which is often boring and sometimes even unpleasant – into a more exciting payment experience. The other approach focuses the added value with the retailer. If the retailer participates, the card has value. If the retailer does not participate, the card has less value. With 'a new way to pay', the retailer wants to be modern and feels the need to accept your card, whereas with 'Carrefour's new loyalty card, Axess', you probably need the retailer much more than he needs you. Whether true or not, that is certainly the basis upon which the negotiations will proceed. 'A new way to pay' will be much easier to sell and negotiate. It will keep you in the driver's seat and will help you keep a higher portion of the added value for your company. In order to protect the margins on your fees and commissions, you need to demonstrate that the lion's portion of the added value is in your payment infrastructure, not the merchant's acceptance (see Table 3.1).

Table 3.1 Two ways to position your payment brand

'Axess, a new way to pay at Carrefour'	*'Carrefour's new loyalty card, Axess'*
No conflict with the retailer's own loyalty card (whether it exists or not).	Requires the retailer to depend on you for his own loyalty card.
Accepted by many retailers *('a new way to pay at Migros, McDonald's, Wal-Mart ...')*	Only accepted within a coalition (and certainly *not* at Migros).
Positions the card's added value within the payment function.	Positions the card's added value within the retailer's participation.
The retailer wants to be modern, feels the need to accept your card.	You need the retailer much more than he needs you!
Easier to sell to retailers.	Very difficult to sell to retailers.

Your marketing agency should be able to provide you with a rich variety of potential brand names that can be used in conjunction with your preferred payment brand, evoking the capacity to surprise and delight customers with instant treats whenever your card is used to pay. For the sake of simplicity, I will use a made-up brand name along with its own catchy little tag line, 'iPay, the smarter way to pay'. The little 'i' in front of 'Pay' can mean intelligent, instant, immediate as well as 'I'.

FOCUS ON PUTTING THE BRAND ON A MAXIMUM NUMBER OF CARDS ... MERCHANTS WILL FOLLOW

Once the iPay brand takes off and it appears that cardholder critical mass is being achieved, merchants will have little reason not to become involved in it. From 2002 to 2003, Axess had grown its cardholder base by 42%, around the same healthy growth as the prior year. The merchant network was still relatively small in 2002, with around 11 000 participating merchants, illustrating the early difficulties Akbank faced when trying to recruit merchants. But something happened in 2003. The perception of cardholder critical mass was achieved. The Axess merchant base exploded, growing by 318% in a single year. In just one year, the Axess merchant network grew to be four times its size the prior year!

Once a merchant places your enhanced payment brand in the store and begins linking surprise incentives to your brand, you have a virtually permanent relationship with that merchant. It becomes difficult for the merchant to back out. As other merchants come on board, you begin to build momentum. Customers prefer your brand because more merchants accept it. Merchants prefer your brand because more of their customers are cardholders. Within nine months after Mashreqbank in Dubai launched its WOW! branded credit card at the beginning of 2004, it had increased its fees and commissions, mostly related to transactions, by 36% and had strengthened its merchant value proposition and pricing policies. In comparison, interest-related revenues increased by 13% during the same period.

Banks that are involved in both acquiring and issuing activities are in a unique position to leverage their merchant network to attract new cardholders and, vice versa, leverage their cardholder base to boost their merchant network. Let's look at several different situations:

1. Banks strong in both issuing and acquiring, for example banks which are among the top three issuers and top three acquirers in their country
2. Banks strong in issuing but weak in acquiring
3. Banks weak in issuing but strong in acquiring
4. Banks weak in both, and finally
5. Banks only involved in issuing.

The suggestions below need to be seen as general ideas which must be adapted for each specific situation.

- *Banks strong in both issuing and acquiring* should begin putting their iPay brand on all the new EMV cards they issue, whether credit or debit cards, in order to immediately establish the perception of imminent critical mass. The primary competitive goal is to prevent the competition from catching up, maintain market share in issuing and acquiring, create stickiness amongst cardholders and merchants, making it more difficult for either categories of customers to defect to other banks, and actually widen the gap in market share with the bank's closest competitors.
- *Banks strong in issuing but weak in acquiring* clearly need to leverage their cardholder base to grow and strengthen their merchant network. In order to recruit merchants for your EMV acquiring services you need to at least create the perception of cardholder critical mass or the perception that critical mass is quickly approaching. If you have a large cardholder base that is being converted to EMV, you will probably be able to recruit merchants through the size of your card base alone. Since merchant acquiring has become such a commodity with most companies competing almost exclusively on pricing, you have an excellent opportunity to create an exciting new value proposition for merchants who would like to have access to real-time

features targeting your large cardholder base. You might be able to break the prevalent commodity rule for acquiring services in your market. This is one of the fastest ways to substantially increase non-interest income. Acquiring does not have to be the commodity it currently is. EMV can help make acquiring more profitable. The first banks to seize this opportunity during EMV migration can secure a greater portion of the added value for a longer period of time.

- *Banks weak in issuing but strong in acquiring* need to leverage their merchant base to quickly recruit more cardholders, and they need to reinforce their leadership position with merchants. Participating merchants can be commissioned to encourage customers to sign up for a Visa or MasterCard iPay credit card, issued by your bank, by placing card application forms in their stores. What other things can you do if your cardholders don't represent a sufficient percentage of a merchant's customer base? Since the majority of your revenues are through your acquiring activities, you may already offer processing services to other banks, such as larger international card issuers that have little or no acquiring activities of their own in your market. You might be in an excellent position to license your iPay acceptance brand to other card issuers who want access to your merchant network. Announcing that one or more card issuers will be joining your network and will be using your brand will make it much easier for you to grow and consolidate your merchant network even further. Your acquiring services will be more attractive to merchants, making it easier to sign new merchants and charge higher fees. Carefully choose which banks can enjoy enhanced EMV features within your merchant network. Charge transaction fees for access to your network, and since cards will be issued with your brand, charge per card fees upon issuance. Use the network to increase your card base only; you can still retain the right to only allow merchants to place application forms for your cards in their stores, so that the merchant network is only used to grow your cardholder base. If the other banks using your brand are large issuers already, this should not be a sticking point in the negotiations.

- *Banks weak in both issuing and acquiring* can choose from a number of different options, depending on whether they have a strong preference to grow either their issuing or their acquiring activities. Without taking into account individual strategic preferences, the easiest opportunity to pursue would probably be to first concentrate on growing the acquiring base. Such banks are probably already focused on retail niches. They can create an 'iPay' brand, put it on all of their cards, and use it to grow and reinforce their position within the purchasing categories they are already strong in. They can explore licensing the brand to other issuers who want access to the bank's particular acquirer niche. This will help create the perception that cardholder critical mass is coming, which in turn will help grow your merchant network faster.

- *Banks only involved in issuing* need to partner with a strong local acquirer. You might want to create an iPay brand, launch it on your cards, and attract the largest acquirer by offering him the exclusive right to acquire iPay card transactions ... plus the right to sublicense iPay to other issuers in a couple of years. Show the acquirer how your strategy will help him remain the dominant acquirer for many years. Although in time your brand will be used by other issuers, you will retain control over how the brand evolves in the long term and you may be able to insist that no other issuer be allowed to recruit cardholders through the iPay merchant network. Plus, you should be able to generate lucrative non-interest revenue from iPay transactions on cards issued by other banks, revenue that is high margin because the actual processing is performed by your acquirer partner.

ENHANCED PAYMENT BRANDS FOLLOW IN THE FOOTSTEPS OF TRADITIONAL PAYMENT BRANDS

Just as merchant acceptance is a fundamental part of all payment network brands today, such as Visa, MasterCard and American Express, it should be clear by now that the merchant network is also a fundamental part of enhanced payment brands like Axess, WOW! or our fictional iPay brand. Banks that have remained active in acquiring will be able to benefit faster than others. As they create and deploy enhanced payment brands, they will be following in the footsteps of traditional payment brands and can learn much from studying how Visa, MasterCard and American Express became part of a very small group of payment brands to succeed worldwide.

Why are there not hundreds of payment acceptance brands around the world? Why are there not 100, or 50, or even 20? Why are there less than a handful of payment brands? Could there be hundreds of iPay brands across the world? Or will there eventually only be a handful? Merchants don't want the hassle of processing dozens of different payment brands, so there is a natural pressure to reduce the number of payment brands within a country to something manageable. Similar pressures will impact enhanced payment brands. Part of the answer can also be found in the fact that, in most countries, there are not more than three or four acquirers processing 80% to 90% of all merchants. Many countries have only two. Some countries have only one. At any one time, a merchant only works with a single acquirer. If several acquirers offer enhanced payment services and all are competing for the merchant's business, the merchant will prefer the acquirer that gives the merchant access to the largest cardholder base. This promotes consolidation around a small number of strong enhanced payment brands within the same country. If each acquirer created an iPay acceptance brand, this means that each country may only have room for at most three or four major enhanced payment brands, with smaller issuers aligning themselves around these brands. In fact, one could argue that the same market pressures that have resulted in just a handful of globally recognized payment brands may also cause just a handful of enhanced payment brands to emerge over time.

We are already beginning to see this happening in Turkey, where two strong enhanced payment brands compete head to head, Akbank's Axess and Garanti's Bonus. In 2003, Garanti licensed its Bonus brand to another card issuer, Denizbank, enabling it to use the brand name and the entire merchant network. Denizbank issued 500 000 Bonus cards in one year and increased its card profits by 55%. Garanti strengthened its merchant value proposition by ensuring that a larger volume of customers would be attracted to Bonus merchants. Garanti is now offering other banks help in operating a bonus pool, designing marketing activities, preparing a scoring and collection system for revolving credit and general help in EMV migration.[30]

This is actually very analogous to the birth of the credit card industry itself. The Visa card had its genesis in the mid-1950s as a California service of the Bank of America called BankAmericard. Concerned with possible erosion of their customers, five California banks jointly launched MasterCharge in 1966. In turn, Bank of America franchised its service. Other large banks quickly launched proprietary cards and offered their own franchises. Action and reaction were soon rampant.

Merchant-acquirers have a particularly vital reason to use EMV migration to create new payment services and new enhanced payment acceptance brands linked to those services. Today, acquirers recruit merchants by negotiating transaction fees. Differentiation based primarily on pricing has created an industry with razor-thin margins. If an acquirer creates a new brand for new EMV payment features, launches it and manages to license the brand to several card issuers, merchant recruitment can become significantly easier. Existing merchant contracts can be made more secure. Negotiations need no longer be based only on pricing, since a whole new discussion on the added value related to the brand suddenly becomes possible. The acquirer can more easily win over new merchants for payment transaction processing when the brand is part of the overall service package offered to merchants.

Once a merchant switches over to a payment acquirer in order to link the merchant's incentive programme to the acquirer's acceptance brand, the merchant becomes dependent on the acquirer for a long period of time. As time goes by, it becomes more and more difficult for the merchant to stop using the brand and hence to switch again to a rival acquirer.

Major banks that are both card issuers and merchant-acquirers will be able to enjoy combined benefits from their acceptance brand. In addition to new acquirer revenues, these organizations can enjoy a higher share of the merchant's transactions and reduced cardholder acquisition costs. Participating merchants will want to encourage their customers to become iPay cardholders and will place 'take-one's' directly at the point of sale. Active merchant participation in recruiting new cardholders is far more effective and less expensive than mass mailings offering enticements and introductory offers of free credit.

How about marketing coalitions like Air Miles? Many of these are already loyalty brand managers and will be able to expand their current services to also cover payment features leveraging the new capabilities of EMV cards. They might do this in alliance with a major acquirer, or they may choose to work with multiple acquirers.

What do companies stand to lose by not owning a dominant enhanced payment brand? Quite simply a very privileged relationship with merchants and cardholders. This is a key element of the overall value proposition offered by EMV. Control of an EMV acceptance brand will prove to be a valuable asset.

WHAT ABOUT A MASTERCARD OR VISA 'iPAY' BRAND?

If I have an EMV card issued in the UK with a MasterCard brand, I know that I can use my card at any merchant in the world that accepts MasterCard branded cards. Today, the payment function is essentially identical whether I use my card in the UK or whether I use it abroad. There is no downgrading of features when I travel. Quite the contrary: many cards offer built-in travel insurance and other features, which actually upgrade the card's capabilities when used abroad. If I enjoy receiving special incentives and other useful information on my credit card receipt when I use my card frequently in the UK, would I not find it strange that my card is somehow downgraded when I travel and that I can no longer receive incentives or information elsewhere? I would prefer a card that promised me the same high level of features and services everywhere.

Very few of the 25 000 Visa and MasterCard member financial institutions are involved in merchant-acquiring. A few hundred financial institutions are responsible for some 80% of all Visa and MasterCard cards worldwide, yet even amongst this smaller group, relatively few are merchant-acquirers. The largest card issuers are sometimes acquirers in a few countries, but often have no acquiring activities in dozens of other countries. What role might Visa and MasterCard play to help financial institutions get the most value out of their EMV cards, wherever their cards are used?

MasterCard offers financial institutions many different consumer payment products and services, each with its own consumer brand. A quick visit to

MasterCard's website reveals all kinds of brands, such as MasterCard™ credit and debit cards, Maestro™ online debit cards, Cirrus™ cards for ATM access, Mondex™ e-cash, PayPass™ contactless cards, MasterCard SecureCode™ for secure online shopping, MasterCard MoneySend™ person-to-person funds transfer service, World MasterCard cards for seasoned travellers, BusinessCard™ cards (and associated services like a Business Bonuses™ programme and a Smart Data OnLine™ financial management service), Corporate Multi Card™ cards, Corporate Fleet™ cards and literally dozens of others. Visa's website provides a similarly large list of Visa branded payment products and services.

Both associations play a fundamental role by creating interoperable payment products which banks can bring to market quickly and easily. A bank that offers Cirrus cards knows that its customers will be able to use their Cirrus card at ATMs around the world. Banks that offer the MasterCard MoneySend service know that their customers will be able to send money to customers of other banks that also offer the service.

Visa and MasterCard have worked together for over a decade to define common chip card specifications ensuring interoperability of basic fraud management features. This is the minimum that all cards and terminals must support. Beyond that, the associations compete aggressively to offer distinctive features that help their member banks get maximum value out of chip cards. Both Visa and MasterCard have created proprietary specifications for storing enhanced EMV data in the chip, called VS3 (Visa Smart Secure Storage) and MODS (MasterCard Open Data Storage). These specifications are one of the building blocks for an iPay brand. Since the underlying technology is already available in virtually all EMV cards and terminals on the market today, creating a consumer brand which can enhance traditional payment brands would not be a complicated next step forward for Visa or MasterCard. Along with the brand, the associations would define distinct rules on how certain features are used by issuers (such as real-time mileage accrual and redemption, cardholder preferences, and so on) and how other features are used by acquirers (such as merchant promotions, welcome gifts and so on).

In this scenario, a card issued with the iPay brand would offer similar functionality whenever it is used at a participating merchant displaying the iPay brand, anywhere in the world. Issuers don't have to be active in acquiring to fully benefit from EMV; any issuer can provide iPay branded cards to customers. Acquirers don't have to offer merchants limited access only to cards issued by the acquirer. Merchants can enjoy the same added value services across all iPay branded cards.

Using the same EMV infrastructure used to combat fraud, an iPay branded enhanced payment service, backed by one of the card associations, could prove to be one of the most powerful ways to resist the strong downward pressure on transaction fees which banks everywhere are experiencing. Why should a bank restrict its EMV infrastructure to features which only combat losses due to fraud, when it could be using the same chip infrastructure to combat the much larger losses due to fee reductions resulting from commoditization? The card associations have a central role to play in helping banks maximize the value of their EMV investment to address commoditization and fraud together, simultaneously.

EMERGING MARKET EXPANSION STRATEGIES

From a financial institution's perspective, private store cards and co-branded cards are zero-sum games. If I win a deal to create a co-branded card for a particular retail chain, my competitors lose. If they win the deal, I lose. The same is true for general-purpose credit cards. If I can get a customer to use my bank's credit card more often, that customer will use other bank cards less often. I win, you lose. You win, I lose.

This was not always the case. Zero-sum games are a characteristic of a mature market. The first companies to issue credit cards started by taking big chunks of market share away from cash and cheques. It was only when the market began approaching a point of saturation that credit card companies were forced to battle to take market share away from each other.

Long-term growth in all markets is initially powered by relatively short bursts of non-zero sum competition, followed by relatively long periods in which the market becomes increasingly mature. The market structure is defined at the early stages. Initial market leaders tend to maintain their position of strength throughout the life of the market, until a new paradigm comes along and makes the old approach obsolete. Competition becomes increasingly difficult as the market gradually becomes a zero-sum game.

Today, EMV credit and debit cards are particularly effective as a means of taking further market share away from cash and checks. The opportunity again exists to take advantage of the upcoming burst of growth that will inevitably happen as financial institutions stake their positions as leading providers of card products in this new market.

An enhanced payment brand like Axess, Bonus, WOW! or iPay is the single most important key to triggering a snowballing effect that will attract merchants, customers and card issuers. As the web of interrelations is woven tighter, its strength grows exponentially. As more merchants come on board, cardholders find the programme more compelling. As more cardholders join and become active participants, merchants find the programme irresistible. As card issuers begin adopting the brand on their cards, other card issuers become attracted to the programme and find they must participate *now* if they are to be part of the entire market's burst of growth.

In zero-sum games you always try to hide your strategy. But in non-zero-sum games you might want to announce your strategy in public so that the other players need to adapt to it. When Gorbachev did away with 10 000 tanks, he announced it clearly and unilaterally put an end to the Cold War by eliminating the West's justification to continue investing heavily in weaponry. Being secretive would have had no effect.

Emerging market expansion strategies are based on co-evolution and non-zero-sum games. During this phase of development, control and secrecy are counterproductive. This is why it is wise to openly license your brand to many other issuers, even if they are your competitors. This will help establish your company as the market leader for a long period of time.

Kevin Coyne, of McKinsey & Co., explored the competitive dynamics of network-based businesses in an article published in the *Harvard Business Review*. He looked specifically at how retail banking established an ATM network in the US. In the late 1970s, Citibank became the first to offer large numbers of ATMs at its branches in New York, providing brand new 24-hour service to customers. Other banks had installed a few ATMs as well, but none was widespread. Citibank's proprietary network gave it an enormous advantage. In response, rivals, led by Chemical Bank, banded together and formed the Plus network. In network-based businesses, smaller players generally succeed when they band together to compete against the dominant provider. Citibank initially declined to join Plus. But as often happens in network wars, the greater combined numbers of the small players overwhelmed the large single player. Although Citibank was the most convenient single bank, it could not match the combined presence of the other banks. That would have required placing a proprietary outlet almost everywhere anyone wanted to do banking. The Plus network was the winner in the battle for perceived ubiquity.[31]

Dee Hock, founder of Visa International, recalls the early days of the credit card industry when banks failed at establishing proprietary merchant acceptance networks:

> One bank issued cards with a hole in the centre and supplied merchants with imprinters which had a matching steel peg in the bed to shatter the cards of competitors. No stupid sharing of point of sale devices for them. Let the merchant worry about irate customers. Another bank, dreaming of riches from rental fees, methodically

removed competitor's imprinters and installed their own, only to be thrown in the slammer by the local sheriff, who happened to be a loyal customer of the competitor.[32]

Creating a network is expensive and risky if it is done in a closed, proprietary manner. Developing a network becomes far less expensive and virtually risk-free if it is shared among competitors. This suggests that the strategy for a strong, established market leader would be to launch a proprietary network to gain time-to-market advantages, and then suddenly open the network up to competitors once they appear to be on the verge of banding together.

SWARMS, LANGUAGES, NETWORK EXTERNALITIES AND THE LAW OF INCREASING RETURNS

Have you ever seen a swarm of bees take off? Kevin Kelly provides a description of the process in his book *Out of Control*.[33] When bees outgrow their hives and feel the need to move to a new place to live, individuals will search out potential sites. A bee will fly back to the hive and dance to communicate what it has just found: 'Go over there, it's really nice.' Other bees watch, and then go out to see for themselves. They come back and say something like 'Yeah, it really is nice'. The intensity of the dance, and the number of bees that concur, encourage more and more bees to go and check the new place out. Multiple sites are championed in this fashion, but eventually critical mass builds for one particular site, the hive becomes more and more excited and finally bursts out, rises into the air, hovering over the empty hive like a soul leaving its body, and then heads out toward the new site.

Languages evolve in a similar manner. Speaking a language shared by many people is more valuable than understanding and speaking a language limited to a few individuals in a remote mountain village. Looking at language evolution and changing cultural influences across continents over several centuries, one can clearly see things that feel similar to the swarming effect that happens when bees move to a new site. A language becomes more and more valuable as more and more people speak it. At one point it achieves critical mass and then increases exponentially, filling up an entire region, country or continent. Once critical mass is achieved, things happen quickly. Two generations is enough to completely switch to the dominating language.

Swarms and languages illustrate what economists call 'network externalities'. If the value of an object increases as more and more people use it, network externalities happen. A single fax machine has no value. Two fax machines make each machine twice as valuable: the size of the network counts far more than the machine itself. Telephones benefit from the same effect; so does the Internet. A network's value grows even faster than the number of members added to it. A 10% increase in customers for a company that does not benefit from increasing returns may increase its revenue by 10%. But a 10% increase in customers for a telephone company could raise revenues by 20% because of the exponentially greater numbers of connections between each member of the network. Most software packages also benefit from this, as do operating systems of course, but also word processing applications, spreadsheets and presentation managers. If people around you all use Word, you are better off also using Word so that you can easily share files, learning and advice.

Payment acceptance brands react to network externalities as well. The market prefers a brand that many customers have on their cards, and that many merchants use for their electronic incentive programmes. As more and more cards carry a particular brand, more and more merchants will want to accept the brand. This in turn makes the brand more valuable to other customers that don't yet have an iPay branded card. They become cardholders. That makes the whole network more valuable to other merchants. They come on board. And so on and so forth. Once critical mass is reached, things accelerate. The iPay network grows exponentially in value and soon becomes irresistible to customers and merchants who appear to be saying, 'I must have that card!'.

Markets that benefit from network externalities abhor niche strategies. In the 1970s and 1980s, a number of industries developed specialized proprietary networks linking buyers and sellers together for order entry, invoicing and inventory management. Various incompatible Electronic Data Interchange (EDI) protocols were defined and put into place at great cost to their champions. Today, most of these have all but disappeared, swallowed up by the mass-market Internet. The remaining EDI companies were able to understand that their intellectual property was not in the area of infrastructure like communications protocols, but in content which is easily transferred to the Internet. Of course, the Internet itself evolved out of a niche market of scientists and university researchers. The point is that with network externalities, the niche phase is never long-lasting. It will always be replaced by a mass deployment phase at some point. Niche markets will always be better addressed by an inexpensive product offering aimed at the mass market.

At the beginning, niche payment brands delivering incentives in niche markets may emerge in some areas. Brands may appear for specific cities, states or regions – something like 'The New York Card', with an apple as part of its logo. Other brands may appear for categories of customers, like France's youth card, Carte Jeunes, a smart card brand dedicated to the under-26 crowd. However, because of network externalities, the market will always prefer, if not demand, a single, widely available enhanced payment brand. A common brand will be made available to a much larger number of cardholders, which will attract a much larger number of merchants. This is inevitable. Just as the Internet instantly swallowed up professional EDI networks the moment it entered the mass market scene, a common enhanced payment brand will prove to be irresistible. This is also why payment brands like Visa and MasterCard are so powerful. It is unthinkable for merchants to have to enter into agreements with hundreds of individual banks in order to process credit cards that only carry the bank's name. New enhanced payment brands will thrive under the same strategy. Because merchants don't sign payment-processing agreements with lots of banks (and don't paste hundreds of bank logos on their doors) merchants will prefer not to sign agreements with lots of individual banks to enjoy enhanced payment features on a maximum number of cards.

It is virtually impossible to fight against network externalities or the law of increasing returns, short of getting governments and cartels to restrict the use of payment cards as marketing vehicles (which is still the case in France today, but this is changing) or to outlaw the practice of promotional marketing incentives in general (as was the case in Germany until recently). Early champions of niche products and services often spend fortunes fighting a losing battle against the law of increasing returns. Early EDI protagonists never recuperated their investments in proprietary protocols that were easily overrun by the Internet. They were forced to transfer their services to the web.

Companies that battle against the law of increasing returns lose money. Companies that encourage network externalities and help a fledgling product category emerge and begin benefiting from the law of increasing returns almost always end up in a dominant market position for many years, at least throughout the life-cycle of that product category.

This represents a tremendous opportunity for companies that are now beginning to champion customer-centric EMV payment features in addition to EMV's fraud prevention features.

THE LAST 50 YEARS AND THE NEXT FIVE

By understanding how payment cards have evolved over the past 50 years, we can begin to see where the market is going over the next few years. It turns out that enhanced payment brands are just a new twist on how the payment card industry has addressed other issues in the past.

Store credit cards were born in the US in the 1940s. These cards were initially accepted for payment at a single large department store or chain and were issued by a single financial institution. Sears, JC Penney and Kmart are examples of organizations that have issued private cards.

The first store cards quickly gave rise to a number of local credit cards issued by large regional banks and accepted across a variety of merchants. In the late 1940s, a number of US banks started issuing their customers scrip – a certificate representing or acknowledging value – that could be used like cash in local shops. The practice was formalized in 1951, when the Franklin National Bank introduced the first modern credit card.

In the 1960s, local credit card issuers joined forces, giving rise to the associations that later became known as Visa and MasterCard. Bank of America was first to extend the credit card throughout the United States by introducing the BankAmericard, which later became Visa, and by franchising a single bank in each major city. Each bank was responsible for recruiting local merchants and enrolling cardholders.

At the same time, loyalty programmes began proliferating. Punch cards are valid at a single merchant location or chain and are not linked to any particular payment method. Co-branded cards issued by a single issuer are valid at many merchant locations for payment, but only provide loyalty rewards at a single merchant location or chain. Private loyalty cards are essentially store credit cards with a loyalty programme attached.

Then programmes became more complex. Single issuers like American Express developed programmes like Membership Rewards, pooling mileage points from many merchants into a common bucket. In some cases, the same reward programme was adopted by multiple card issuers, giving rise to multi-issuer loyalty brands like Air Miles (initially launched by British Airways) and UK's Smart programme (initially launched by Shell).

Let's take a moment and look at what's been happening here. During the market's early phase, we can identify two major product evolutions over a 15-year period (store credit cards and local credit cards) … and four major product evolutions over the next ten-year period (card associations, store loyalty cards, co-branded cards and mileage programmes). Things are accelerating.

EMV migration provides an opportunity to create more powerful card products and brands. By applying the market rule of concurrently addressing the needs of banks, merchants and customers, and by extrapolating what has been happening in the payment card industry over the last 50 years, we can develop a vision of where the market is heading.

We're already beginning to see the first single issuer EMV cards that let multiple merchants each trigger their own incentives based on behaviour data stored within the same smart payment card. This is a recombination of simple credit cards (like those issued prior to the existence of Visa and MasterCard), with punch cards and paper coupons that already existed long before credit cards were invented. As such, it represents an evolution that uses new technology to create new combinations of proven products and services, each with their own established business model.

We're also beginning to see the next step in the process, with the creation of enhanced payment brands used by multiple card issuers so merchants can trigger incentives across a large number of cards. Given the payment industry's long-standing interest in uniting around a few common payment brands, the industry will soon be moving towards a few common enhanced payment brands.

CREATING COMPETITIVE ADVANTAGE THROUGH BETTER BRANDING – FOUR KEY OBSERVATIONS

1. Customers decide which cards earn the privilege of residing in their wallets based on the services and features offered by a particular brand, not the card's technology. The traditional payment card market is already mature. Branding issues have long been resolved. The new EMV smart card offers a richer payment experience and will require its own distinct branding strategy. An enhanced payment brand is required, something like iPay so cardholders and merchants easily know where the cards can be used and which customers are entitled to the enhanced payment features offered at the merchant's point of sale.
2. Once a merchant places your iPay brand in the store and begins triggering incentives linked to that brand, you have a virtually permanent relationship with that merchant. It becomes difficult for the merchant to switch to another provider. As more merchants come on board, cardholders find the programme more compelling. As more cardholders join and become active participants, merchants find the brand irresistible. As card issuers begin adopting the brand on their cards, other card issuers become attracted to the programme and find they must participate *now* if they are to be part of the entire market's burst of growth.
3. Just as merchant acceptance is a fundamental part of all payment network brands today, such as Visa, MasterCard and American Express, the merchant network is also a fundamental part of enhanced payment brands like Axess, WOW! or our fictional iPay brand. Banks that have remained active in acquiring will be able to benefit faster than others. As they create and deploy enhanced payment brands, they will be following in the footsteps of traditional payment brands and can learn much from studying how Visa, MasterCard and American Express became part of a very small group of payment brands to succeed worldwide.

4. Initial market leaders tend to maintain their position of strength throughout the life of the market, until a new paradigm comes along and makes the old approach obsolete. The opportunity again exists to take advantage of the next burst of growth as the payment function becomes richer and offers new customer service and marketing opportunities. Emerging market expansion strategies are based on co-evolution and non-zero sum games. During this phase of development it is wise to openly license your brand to many other issuers. This will help establish your company as the market leader for a long period of time.

CHAPTER 4

Creating Competitive Advantage Using Technology

In 1979, with the improvement of electronic processing, electronic dial-up terminals and magnetic strips on the back of credit cards allowed retailers to swipe the customer's credit card through the dial-up terminal, which accessed issuing bank cardholder information. The advantage of this system, besides saving paper, was the increased speed of processing authorizations – one to two minutes. It also decreased credit card fraud.

Gareth Marples[34]

In the past, payment systems concentrated intelligence centrally, on large mainframe computers that could respond instantly to hundreds of thousands of payment authorization requests. The merchant's terminal was a simple, inexpensive device that had little or no intelligence. It didn't need to be intelligent. All it had to do was faithfully read the magnetic strip on the back of the card, understand the amount keyed in, dial up to its host, perform handshaking, transmit the purchase amount along with whatever was stored in the customer's magnetic strip, and wait for an authorization or a refusal. That's it. The terminal application was simple to write and maintain.

Banks provided their payment application specifications to terminal vendors, as part of their technical requirements document, and vendors modified the terminal's software to take into account the bank's requirements. Each terminal vendor rewrote the payment software for each model offered, and built the software cost into the price of the hardware, essentially providing the software pretty much for free, as there wasn't really very much to justify charging for it. Because terminal manufacturers typically offered several distinct models of terminals, each with a unique operating system and development tools, the software had to be rewritten for each of those. This might seem like a waste of energy, but the application was so simple that it was easier to rewrite it than to adopt a more modern, portable application software architecture.

Then smart cards came along, with the ability to do much more processing at the point of sale, often with no connection to the host system at the moment of purchase. Suddenly the terminal had to go from relatively dumb to relatively smart. In one jump. Today, we are in the middle of a major transition. Over the past few years, some of the world's leading point of sale terminal manufacturers have had lots of trouble moving to the new mode of thinking based on significantly increased intelligence in the card and in the terminal. Some have even openly resisted the move to smart cards and a few still offer terminals that are quite basic.

Fundamental changes in the marketplace are of course a dangerous moment for the incumbent players. Since the underlying philosophy of payment terminals is going through such profound change, manufacturers cannot take their existing machines and throw in intelligence, which is precisely what many companies initially tried to do. Attempting to layer intelligence onto a product that can't adequately support it leaves the field wide open for newcomers with new products, redesigned inside out for new market requirements. As terminals go from being dumb to being smart, the current market leaders have had to depend and capitalize more on their sales people's personal relationships with customers, and less on the company's technical products – at least until their engineers had the time to catch up and create new products that responded adequately to the market's requirements. This is what Verifone and Hypercom, the number one and number two vendors at the time, had to do, while Ingenico, a French terminal company that began offering smart card terminals to French banks at least a decade before the incumbent players, was able to leverage its early market advantage and become a formidable opponent. Today, both Verifone and Hypercom have caught up, but not before giving Ingenico lots of time to carve out a substantial share of the worldwide terminal market and take the leadership position. In terms of revenue, Ingenico seemed to clearly be in the lead in 2003, with sales of €356 million (US$434 million). Hypercom, like Ingenico, is publicly traded and reported a 2003 revenue of US$231.5 million. Privately-held Verifone did not report its 2003 revenue, but estimated that its sales for the 12 months ending October 2003 would reach US$335 million.[35]

According to Verifone:

> Only a decade ago, a merchant could accept a single type of credit card and satisfy most customers. That has changed dramatically with the proliferation of card-based payment and value-added options in the highly competitive retail marketplace. Today, many customers carry a variety of cards, and they expect retailers to readily accept whichever card they choose for a given purchase. To compete effectively in this environment, merchants have offered a growing variety of credit and debit services, electronic stored value cards and loyalty programs designed to encourage customers to patronize their stores. This has added significantly to the complexity of today's payment environment. Applications today are becoming much larger, more complex and difficult to manage. Further, newer applications such as loyalty schemes store more data and often must be updated more frequently.[36]

Here is how Hypercom sees things:

> The evolution from magnetic strip to smart card technology has increased transaction security requirements. Also, the number of applications has raised the complexity of point of sale terminals. Unfortunately, many terminal manufacturers have not yet discovered how to deliver easy-to-use merchant and consumer appliances that are smart card-capable. Smart cards now allow consumers some measure of choice at the point of service. They also allow card issuers to customize the behavior of a specific card for a given consumer. To accommodate these complexities, POS terminals require improved user interfaces and more CPU power through single processors or parallel multi-processing. The ultimate selection criterion, however, remains fast overall transaction times.[37]

Both Hypercom and Verifone now see the market in a manner quite similar to Ingenico's. All now see increasing complexity within the terminal as the source of substantial added value. Ingenico's chairman, Jean-Jacques Poutrel, is not too concerned about the danger of equipment prices plummeting. 'Given the increasing complexity of software and strong customer demand for greater transaction security, plain-vanilla terminals are doomed to rapid obsolescence,' he says.[38]

MARKET PRESSURES PROMOTE GREATER TERMINAL COMPLEXITY AND DIFFERENTIATION, NOT LESS

Terminal manufacturers increasingly compete on the form factor, that is the display, the keyboard, the printer, the physical footprint, wireless capabilities (so a waiter doesn't have to set the terminal down when a transaction has to be performed online) and other ergonomic matters. There is a growing need to offer competitive differentiation across all of these features. Terminal manufacturers also compete on the operating system, the software architecture and the software development tools. No two competing suppliers use the same internals.

Just looking at three popular terminal platforms, the Ingenico Elite 510, the Verifone Omni 3750 and the Hypercom ICE 5500, we can list dozens of substantial differences. One terminal uses a 16-bit processor, the second uses a 32-bit processor while the third uses an 8-bit processor. Two platforms offer a multi-application capability, meaning that different applications each run with a distinct executable code, while the third only supports a single executable application in the terminal. One terminal supports a proprietary limited type of multi-tasking, another supports full multi-tasking, while the third supports no multi-tasking. Only one of the three terminal platforms supports fully addressable memory in a modern fashion. The other two platforms require the programmer to manage each page of memory separately. If new code is added for a new feature, the programmer often has to move code around all over the place to make pieces fit into each page of memory correctly. Each terminal has a completely different display format: four lines of 16 characters, or eight lines of 21 characters, or a fully addressable touch screen with 160 × 80 pixels. Even modem speeds are different: two still have archaic 1200 baud modems, while the third offers a more modern (yet already getting old) 14 400 baud speed.

Because each terminal's hardware, software architecture and even payment applications are all so different, each platform has its own peculiar way of handling remote maintenance functions such as updating the software without having to visit the site, downloading new terminal parameters and performing remote diagnostics of deployed terminals. Each terminal vendor offers a distinct, proprietary terminal management system for this purpose. Banks that support several different terminal platforms are obliged to also support several distinct terminal management systems, one for each platform. Often it's cheaper for the bank to only support one brand of terminal.

The worst part of all this is that competitive pressures are creating more differentiation, not less! There is no incentive for terminal manufacturers to work towards a common platform. On the contrary. To sell more, better, and at higher margins, they must differentiate. They have no choice. On a more positive note, one must assume that the market somehow benefits from so much differentiation in such small packages. And, in fact, the market does. The terminals available today are far more powerful than those of a few years ago. And on top of that, they cost less.

Differentiation is not limited to the terminal hardware and software architecture. Differentiation also extends to the actual payment application in the terminal, since each terminal manufacturer tends to create applications only for one terminal platform ... his. Terminal manufacturers subcontract application development to third-party software developers, but since the terminal manufacturer is the developer's immediate customer, the one paying, the software developer tends to make the application work on only one terminal platform. Because of this, each vendor's payment application has a unique look and feel. The same payment function, whether a simple credit or debit function, or a more advanced payment function including gift cards, for example, can appear extremely different on each platform. Each platform displays different messages, all for the same function. There are many ways to start a transaction, such as 'Insert card', 'Card not inserted' or 'Swipe/insert card'. There are many ways to alert the operator to a problem with the terminal, such as 'Paper jam', or 'Paper ERR: Wait' or even cryptic messages like 'Error 200'.

All of these differences naturally result in high training and support costs for banks that want to use more than one terminal platform. The differences act as a lock-in mechanism, making it too expensive for a bank to use more than one platform. Banks that want multiple suppliers have to go through multiple redundant efforts with each supplier. They have to work with each supplier individually to explain their technical and functional requirements, often answering the same questions over and over again. The actual development is performed separately by each terminal supplier (in reality the supplier's designated software developer). Each platform must be tested and certified. The bank's point of sale deployment team must be trained on each platform. Merchant helpdesk personnel must be trained and common documentation must be written by bank personnel. The bank must go through this process over and over again for each and every modification as well. Y2K, the Euro, EMV ... all required each bank to go through the process separately, with each individual supplier. The bank has to go through the process after mergers and acquisitions as well, in order to combine multiple authorization platforms for example. The current approach to terminal management results in high costs to the bank, long time-to-market on new features, and low flexibility due to the substantial hassles and headaches involved.

What has happened here? Why has terminal management become such a hassle? It wasn't so unbearable, before. Why has it become such a source of headaches all of a sudden?

The current approach worked well in the past but it is no longer suited for the new environment driven by EMV and greater intelligence at the point of sale. The merchant processing industry is still using the same old techniques developed for the prior generations of magnetic strip payment systems. Terminal management activities need to be adapted to reflect the new reality of more complex payment systems with real-time intelligence built into the payment terminal.

There is a simple answer to the problem. Today, banks provide their payment application specifications to terminal vendors, who each then subcontract development to a terminal software company, which generally only develops software on one or at most a couple terminal platforms. The result is that the software application is completely different for each terminal. The solution is simple. Instead of providing their specifications to multiple terminal vendors, all banks need to do is provide their specifications directly to a software developer who will develop the same application for all the terminals that the bank wants to support. When the software developer works directly for the bank, as opposed to working for the terminal manufacturer, the software developer's interests instantly become well aligned with the bank's interests.

TERMINAL SOFTWARE DESIGN IS A NEW BUSINESS, INDEPENDENT OF TERMINAL HARDWARE DESIGN AND MANUFACTURING

When terminals didn't need to know very much it was easy to write software for them. The payment associations and merchant-acquirers provided specifications on their payment authorization protocols. These very basic protocols were programmed by a terminal manufacturer's engineers and were loaded into the terminal's firmware. Terminal software was simple to write and easy to maintain, since all the heavy duty work was being done at the host.

Today, terminal software is several orders of magnitude more complex than in the past. Payment software capable of performing an offline authorization does not even resemble prior generations of terminal applications. Nor does it resemble some kind of simplified version of the host application. You cannot take an old credit card application and add lots of intelligence to it. Nor can you take a host application and somehow stuff it into a tiny payment terminal. You can't even take some of the basic authorization algorithms from the host and reuse them, as the authorization algorithms are totally new and different. Nor is the new generation payment application a hybrid between an old terminal application and an old host application. It is a completely new breed of software. Not only must it address fraud issues in compliance with EMV mandates, it also must enhance the payment function with customer-centric EMV features which were unavailable in the past using older magnetic strip technology.

This fundamental change in the nature of terminal software has initiated a change in the value chain. The complexity of the application has given rise to specialist software companies that are independent of terminal hardware vendors. This is a consequence of the terminal industry's maturity. Just as hardware and software in the computer industry has been separated and companies must specialize in one or the other to succeed – and not both – terminal hardware companies and software application companies are distinct entities that must each specialize in their particular area. Dell chose to specialize in hardware, Microsoft chose software. Both companies show no signs of problems with their respective identities. Wall Street recognizes this and rewards it.

Most terminal manufacturers today recognize this. This is why most now provide software development kits for third-party developers. Unlike terminal manufacturers, software companies can develop their applications to be portable across terminal platforms that are not at all compatible. The terminal manufacturer has no incentive to do this, whereas the software vendor does. By making the application available on a larger number of platforms, the software vendor can sell the application to a larger number of customers (see Table 4.1). Customers can mix hardware platforms from various manufacturers and still use the same application.

Table 4.1 The old way and the new way of developing POS software

The old way of developing POS software	*The new way of developing POS software*
An acquirer provides requirements and specifications to each terminal vendor, then works in parallel with each vendor to provide guidance, support and other resources that the vendor needs in order to deliver the terminal product required by the acquirer. Multiple redundant efforts result in higher costs to the acquirer.	The acquirer works once with an independent software vendor to design and develop their specific terminal application. The acquirer informs terminal vendors of the requirement to support the software vendor's products.
Each terminal vendor assigns a team of engineers to develop payment applications for their terminal. This cost is built into the price of the terminal.	The independent software vendor's engineers develop a single application that runs on the terminal platforms chosen by the acquirer.
Custom requirements must be redeveloped by each vendor. Because vendors are not all available at the same time, deployment of new features is delayed until the final vendor is done.	New customer requests are developed once, and then ported to multiple terminals.
A different user interface for each platform results in excessive training of the acquirer's helpdesk personnel and higher costs and effort to ensure quality.	Because the user interface is the same across all platforms, merchant helpdesk support is easier to manage, more cost-effective and of higher quality.
A distinct user guide is required for each application on each terminal. The user guide for an EMV co-branded credit card application must exist in multiple versions, one for each terminal that the application supports.	User guides for the same payment application are virtually identical across all terminals. Transaction flow, messages, receipt formats and so on are the same across all platforms.

Each terminal has its own, proprietary operating system. Some manufacturers even offer a different operating system for each product line. More and more terminals are programmable in C (a standard computing language), but that's still not always the case. Various flavours of C exist. Each terminal manages peripherals in a completely different way. Printers, displays, modems, card readers, internal clocks are all addressed differently. Memory paging is never done the same way. It took Welcome Real-time over 18 months just to define low-level drivers that were relatively universal. This was necessary in order to create a platform on which to build the payment application in a portable fashion. Moving to a new terminal essentially requires re-writing the low-level terminal drivers and then porting the higher level application. Although this is new in the payment terminal environment, it has been standard software design practice in many industries for a long time. It makes no sense to have terminal manufacturers rewrite payment applications based on detailed technical specifications which are becoming more and more complex. It is far easier to write the application once and to port it to multiple platforms. Maintenance becomes much easier as well. Bugs are fixed in a master version, which is then simply recompiled for each platform.

Banks and merchants are increasingly demanding that terminals integrate numerous standards and norms that ensure interoperability. These include card standards of course – like ISO, EMV, Java Card, Multos, and so on – but also general computing standards like the TCP/IP communications protocol used by the Internet, and standard languages like C and Java. Many of these standards are far more sophisticated than the simple protocols that payment terminals had to support in the past. Time and time again, in the mainframe industry, then again in the minicomputer and PC industries, the widespread adoption of mainstream standards has gone hand in hand with the creation of a strong, dynamic software industry separate and distinct from the hardware industry. In general, independent software companies are best positioned to develop and integrate mainstream software standards into various hardware platforms.

EMV MIGRATION IS AN OPPORTUNITY FOR BANKS TO IMPROVE AND MODERNIZE THEIR TERMINAL MANAGEMENT ACTIVITIES

A bank migrating its terminal infrastructure to EMV, using the traditional approach to terminal application management with two or three terminal suppliers, will face high costs due to redundant efforts in requirements definition, testing, certification and re-certification, training and retraining, helpdesk support, upgrades, modifications and so on. Using the new approach to terminal management, with common applications running across multiple terminal platforms, the labour cost of upgrading to EMV could be cut by at least 50% to support the same two or three terminal platforms. The bank's labour costs could be cut by far more if more than three terminal platforms must be supported.

How many of your current POS-related activities are possibly redundant and might be eliminated? Beyond the first certification with a payment application designed to run on multiple terminal platforms, how much of your quality assurance (QA) and training efforts are possibly redundant? How much is spent verifying essentially the same application, over and over again, for each hardware platform? How much is spent on in-depth training on new terminals? How much can you save on annual certification costs?

How would a common POS application improve your helpdesk speed, quality and accuracy? How many people are dedicated to specific terminals? How frequently do you retrain personnel? How often do people need to research answers? When a merchant calls up asking what 'ERROR 200' means, how often do support personnel need to say, 'Let me call you back', so they have time to look up an arcane message on a terminal platform that is no longer used very often? How much can you reduce the average cost per call?

When deploying a feature or modification across multiple platforms, how much of your effort is possibly redundant? How much time do you spend with terminal vendors individually, on everything from feature definition to recertification?

Elimination of redundant efforts is already an excellent benefit ... competitive advantage due to faster time-to-market is even bigger. How much more business can be won each month if you offer a new terminal more powerful and cheaper than your competitor, who is still in certification? You will be less tied to existing terminals, and better able to quickly offer lower-cost, new-generation terminals soon after they become available. If you have not adopted the new approach to terminal management, but your direct competitor has, how much new business can be lost while your competitor has a better POS product, and you are still in certification?

How much can be saved each month by quickly deploying a modification that improves operational efficiency? For example, how much can you save each month by switching all of your merchants to a single authorization platform, if you have ended up with two or more platforms due to multiple mergers and acquisitions? How much can be gained each month by quickly deploying new customer-centric payment features?

WHEN WILL TERMINALS ALL USE THE SAME OPERATING SYSTEM?

The terminal industry has been working on common specifications for an architecture called STIP – Small Terminal Interoperability Platform. STIP is to payment terminals what Java is to smart cards. In fact, the STIP Consortium actually had its genesis in the Java Card Forum. STIP's mission is to create specifications for a common terminal software platform that supports multiple applications on a terminal and provides interoperability such that an application written for one STIP-compliant device will also run on another STIP-compliant device, whether the device is a traditional payment terminal or some other small, smart card-accepting device with limited resources. The consortium includes many of the major terminal manufacturers, such as Ingenico, Hypercom, Verifone, Sagem, Thales and Axalto, as well as providers of other devices which could become card-accepting devices as well, such as mobile phone operators.

The STIP Consortium presents a compelling vision:

> Imagine four devices, a vending machine, a low-end desktop POS terminal, a high-end hand-held portable terminal with large touch-screen display, a PC (with attached peripherals as required). And imagine three STIP applications, an e-purse application, a loyalty application, a security/ID application. Also imagine the following. Each application was written by a different developer and certified once by the organization responsible for the overall card based system. You are able to load and run any combination of the three applications on any of the four devices without modification or re-certification. You are able to do this using the existing download system for the particular device. The desktop and portable terminals have legacy credit/debit applications already installed which continue to run. The applications automatically adapt to the different display, keypad, and printer configurations. Further, the applications only provide the functionality appropriate for the type of device. For example, in a vending machine, the e-purse application might only do a simple debit based on the cost of the item dispensed, whereas the application may provide re-value on a desktop terminal. The ultimate goal of the STIP Consortium is to make the above scenario a reality![39]

Although the STIP specifications were first released in 2000, none of the major terminal vendors offers or promotes STIP-compliant terminals yet. It is possible that the specifications are still going through important changes from one version to the next. The most recent version of the STIP specifications was only released in 2004. Some terminal manufacturers plan to offer STIP on their next generation of high-end terminals based on 32-bit processors and larger memory. However, in time, each manufacturer's most basic terminals will eventually support open software architectures and the market will eventually adopt such an approach. Industry initiatives like STIP will certainly simplify software development efforts across multiple terminal platforms. But will this alone solve the problems facing banks?

If banks continue to provide their software application requirements to terminal vendors, and continue to rely on terminal vendors for payment software, the problem will still be pretty much the same as it is today, even if all terminal platforms supported the same operating system or the same layered software architecture. Each hardware vendor will still naturally want to develop applications which highlight terminal features that are specific to that vendor's platform, and will therefore naturally tend to create applications that work better on his platform. The problem would still need to be solved in the same way that it is being solved today, without a common operating system.

Banks need to work directly with independent software developers, rather than going through hardware vendors. Terminal software developers need to develop payment terminal applications directly for the banks that use them, as opposed to developing them for terminal manufacturers.

WHEN WILL POINT OF SALE TERMINALS BECOME PC-BASED?

This still appears very far away, although it is quite a common question.

In the early 1980s the high-end cash register market went through a similar mutation, where PCs became the heart of supermarket scanning terminals. At the time, manufacturers of firmware-based, proprietary cash registers all complained in unison that PCs would never be sufficiently fast, robust or inexpensive. Cash registers required dedicated real-time operating systems that had to be programmed in firmware.

PCs are notorious for crashing at precisely the wrong time. Losing the price look-up server during a rush period is disastrous. This was solved by using dual server architectures. PCs are slow to boot up, typically taking up to a minute before they are ready to run, whereas point of sale scanning terminals have to boot up in several seconds. This was also solved by putting the boot process in firmware. Numerous other seemingly unsolvable barriers were cleared in a few short years, resulting in PC-based scanning terminals that were as robust as firmware cash registers, less expensive, more flexible and considerably more powerful.

In less than ten years firmware cash registers had all but disappeared from supermarkets and department stores. They had been replaced by PC-based systems. The most successful suppliers concentrated on building open systems and encouraged the development of an independent software industry that could build innovative point of sale applications that could run on multiple hardware platforms.

Intelligence in the cash register meant companies like Wal-Mart could implement things like just-in-time inventory management right at the cash register, instantly taking into account scanning data as items are actually being purchased. More recently, Wal-Mart has even linked their scanning terminals directly into Procter & Gamble's ordering system, so products can be delivered to individual stores when they are needed. Fashion retailer Benetton claims that they no longer dye their sweaters in bulk. They monitor detailed scanning data sent up by each point of sale scanning terminal every night. This allows them to immediately manufacture the exact colours and styles that customers happen to be buying right now. Wal-Mart, Benetton and others have often attributed their fast growth to the extensive use of technology, especially technology at the point of sale.

Increased intelligence at the supermarket checkout has been an important profit engine for many retailers for quite some time. It was a major source of differentiation for early adopters and has forced late adopters to get into the game and mutate their retail businesses in order to keep up. In 1988, Kmart had around 30% of the discount retail market and Wal-Mart 5 –6%. Just a decade later, Wal-Mart had about 45% while Kmart was down to under 12%. This upset was completely unexpected. Wall Street analysts in the early 1970s were all predicting that Kmart was going to be the dominant retailer of the latter part of the twentieth century, because it was expected to crush Sears and be number one for a long time to come. In reality what actually happened, of course, was that Kmart and Sears were eventually forced to merge as both companies struggled to stay alive. Point of sale technology played an important role in helping Wal-Mart, a relatively unknown retailer, to race by and grab market share away from very powerful incumbents.

Sometimes, speaking with payment terminal suppliers, I can hear echoes of discussions I had over 20 years ago in the point of sale scanning terminal industry, with specialized device manufacturers resisting the move to open systems. History repeats itself. The same revolution that swept through the point of sale scanning industry 20 years ago will in time make its way through the payment terminal industry.

A PROLIFERATION OF NEW DEVICES

Several companies manufacture small, keyring-sized smart card readers with a little one- or two-line display that can show the card's e-purse balance, cash back, mileage, points and last ten transactions. They are designed as promotional objects that can be customized with different colours, logos and even physical designs. Imagine a McDonald's hamburger-shaped keyring reader thrown into a special McSurprise Happy Meal. Every time you use your credit or debit card at McDonald's, your child takes the card from you and inserts it in the reader to see if she has won a McSurprise. The display is very small and limited (therefore cheap) but it can still be effective. The display can begin with 'M---r---s--', then as you spend more and more at McDonald's, the display fills up: 'Mc--rpr-s--', then 'Mc--rprise-' and finally, 'McSurprise!'.

Most card issuers have websites that let customers obtain information about their credit card, pay bills and transfer money between accounts. As more and more PCs become equipped with smart card readers, these same websites will provide additional capabilities. Customers will be able to insert their card and see a graphical representation of the many different merchant offers stored in the chip, presented exactly as they might appear if they were traditional paper offers, for example with holes punched on each of the visits that the customer has made towards receiving a free product. Offers that are about to expire, or for which only one more visit is required to qualify for a free item or service, may appear at the top of the list.

When will most PCs be shipped with a built-in reader? When will my Palm Pilot have a smart card reader in it? I'd like to put my chip credit card in and see the status of my electronic incentives and coupons. When will my mobile phone have another reader in addition to the SIM card reader inside?

As a merchant, it would be amazing to be able to recognize my most valuable customers as they walk into the store. Opening the door and entering a store already causes a bell to go off. What if the merchant were able to change the tone of the ring based on the type of customer that has just walked in? Technically, this will soon be feasible, with contactless cards in people's wallets and a reader built into the door. However, regulatory requirements concerning privacy issues might make it less useful in some markets than in others.

Privacy issues melt away if the customer participates knowingly in the process. I want to be recognized as a valuable customer as much as possible, and be treated accordingly. Say a merchant tells me that once I've spent a total of $1000 in his store, I automatically get upgraded to gold status and get lots of extra services. Essentially, the software allows a merchant to bump customers up in status by providing them with virtual gold cards, storing the information in the chip rather than having to physically change cards. When I get in the queue at the deli counter and take a number, it's always upsetting to be holding number 52 when the sign says 'Serving 28'. What if once I reach gold customer status, the ticket machine recognizes me as a very valuable customer through a radio frequency (RF) device that can read my card as it sits in my wallet? Then it could display 'Serving 52' after customer 28 and before customer 29. In France, the land of '*Liberté, Egalité, Fraternité*', where people may get upset at seeing their neighbour treated with special privileges for something as democratic as waiting in a queue, the system might need to be developed in a slightly different way to hide the fact that some people are being treated differently – for example, by normally giving out even numbers, so there is always room to slide in an odd numbered customer without making the other customers too upset.

All of these capabilities will become available thanks to innovations in smart cards, terminals, RF devices, barcode scanners and personal digital assistants. The convergence of payment methods, smart card technology and customer relationship marketing is the driving force behind a whole new industry.

CREATING COMPETITIVE ADVANTAGE USING TECHNOLOGY – THREE KEY OBSERVATIONS

1. In the past, the merchant's terminal was a simple, inexpensive device that had little or no intelligence. The terminal application was very simple to write and maintain. Then smart cards came along, triggering a move to decentralized processing. Suddenly the terminal had to go from relatively dumb to relatively smart. Today, terminal software is several orders of magnitude more complex than in the past.

2. Just as hardware and software in the rest of the computer industry has been separated and companies must specialize in one or the other to succeed – and not both – terminal hardware companies and software application companies are distinct entities that each must specialize in their particular area.

3. In the past, banks provided their payment specifications to terminal manufacturers, who created unique applications for their terminal platforms, each with a completely different look and feel. This resulted in substantial costs to banks, in terms of multiple redundant efforts in certification, support, maintenance and training. The solution for banks is simply to contract directly with independent software application developers who are able to create applications which function on many different terminal platforms.

CHAPTER 5

Intelligent Money – A Glimpse of the Future

Form has an affinity for expense, while substance has an affinity for income.

Dee Hock, Visa founder[40]

The age-old exchange between buyer and seller, culminating in the transfer of money from one to the other, has gone through a long evolutionary process to arrive at today's complex system. Taking the long view, the moment of purchase has been under constant evolution ever since people began exchanging valuables. Going back in time, far back, long before even the first non-barter payment methods were invented thousands of years ago, people were already negotiating payment based on volume purchasing. Buy more of my product today and you can pay at a lower price per unit. A bird in the hand is worth two in the bush. Promotional incentives at the moment of purchase go back to the dawn of human civilization. Incentives that encourage volume purchasing have always been so tightly linked to payment that the two are indistinguishable.

Information technology is the latest element to enter the evolutionary process happening at the moment of purchase. This began over 25 years ago. Now, the smart card, as a powerful agent of information technology, promises to greatly accelerate the transformation of even the substance of money.

Once information technology was introduced into the buyer/seller interaction, the evolutionary process accelerated rapidly. An explosion of techniques has allowed the merchant to better manage the selling process through point of sale checkout equipment, barcode scanners and online just-in-time inventory management. Banks brought significant improvement to the payment process by placing magnetic strips on the backs of credit cards and installing terminals that could automatically request credit approvals and record the transactions. This was all accomplished over a 25-year period, a nanosecond in the history of buyer/seller transactions.

Recent innovations suggest that this is just the beginning. The industry is continuously moving forward, accelerating relentlessly to further integrate computers and communications technologies, coming together seamlessly at the moment of purchase, an event that often lasts only a few seconds. Smart cards are entering into this complex moment, promising further improvements in payment processing. Improvements like better fraud control, faster transactions, reduced processing costs. Smart cards also promise to deliver new benefits that will improve the relationship between merchants and their customers and combat commoditization by making the payment experience more fun, exciting and memorable. All in that short yet important moment that the buyer and seller finally conclude their transaction. The moment of purchase is certainly becoming increasingly complex in everything that happens behind the scenes, but it is also becoming simpler for cashiers and customers. And the moment of purchase is happening faster than ever before. Punch in an amount, insert the card, type a PIN code, and in several seconds the transaction is over, while behind the scenes hundreds of complex operations were just performed and thousands of others were triggered.

The most profound technologies are those that disappear because, paradoxically, they have achieved a level of complexity that allows them to disappear. Antoine de Saint Exupery once wrote, 'Perfection is achieved, not when there is nothing left to add, but when there is nothing left to take away.'[41] Automobiles at the start of the twentieth century were exclusively for mechanical engineers and tinkerers. Each model had its own specific quirks and the technology had constantly to be supervised and tweaked in order to keep the machine running. Today, automobiles are several orders of magnitude more complex, yet they are so simple to drive that many people have no idea what type of engine is under the hood. Take the ubiquitous technology of alphabets. Writing was originally an elite capability, reserved for government bureaucrats. Today, we are no longer aware of performing what was once considered to be a highly complex and almost magical feat as we constantly read words everywhere, even pasted in the form of brands on goods we see and use everyday. Increasing complexity, power and efficiency becomes irresistible when combined with significant improvements in ease of use. All of our tools have progressively evolved in this direction ever since we began making tools.

The substance of money has not changed very often throughout human history. Money has evolved from the literal (gold and silver coins) to the representational (printed paper money and cheques) to the virtual (electronically processed transactions). Changes in the substance of money coincide with substantial changes in society, culture and government. Moving to a new type of value exchange, like going from gold and silver coins to printed paper money, requires deep, widespread consensus. The expected benefits of a new payment method must be clear, easy to understand, and represent a substantial improvement over prior methods. If the new method is little more than a change in *form*, rather than a change in *substance*, it is much more difficult to get it to take hold. Lacking widespread consensus, it could disappear or, at best, be relegated to some specialized use at the fringes of mainstream commerce. Changes in form tend to be expensive and risky for their champions, whereas changes in substance tend to generate substantial new income streams.

Electronic – or virtual – money already exists. Money became virtual once banks began communicating transaction information through computers, almost 40 years ago. In a relatively short period of time, electronic money became a permanent part of mainstream commerce. Taking the long view, the arrival of smart cards is a continuation of the trend that began when banks first placed magnetic strips on their credit cards. Money has already undergone three stages of fundamental transformation, from literal, to representational, to virtual. Today, smart cards are poised to trigger the fourth stage of substantial change. Money is about to become intelligent, aware of how and where it has been used in the past and adapting itself to its environment.

Ten thousand years ago, as hunter-gatherer tribes began settling down in a farming lifestyle, they were already familiar with the only payment method available. Money was whatever they grew, hunted, or made. Nearly three thousand years ago, the first coins appeared. This was a great leap forward in terms of portability and convenience. Three hundred years ago, printed paper money appeared. Soon after that, cheques were invented. The newly discovered ability to move money readily and access it without carrying it paved the way for the use of plastic cards for payment.

American Express was founded in 1850 as a freight express business. In 1882, in response to the postal money order which was causing demand for the company's cash shipping services to decline, American Express began offering its own 'Express Money Order' service. This was an immediate success. The company jumped on the opportunity and began selling the product at railroad stations and general stores nationwide. American Express had just taken a step in the direction of becoming a financial services company.

Several years later, American Express president J. C. Fargo was on vacation in Europe, where he discovered the difficulty of converting letters of credit into cash. On his return, the company created an elegant solution which only required a signature upon purchase and a countersignature upon redemption. This quickly became known around the world as the 'American Express Travellers Check'. The travellers cheque presented an unexpected bonus for American Express. In between the time when a cheque was purchased and the moment when it was redeemed, many weeks and even months would go by. This created a comfortable cash cushion. The company sold more cheques each month than it redeemed. American Express had inadvertently stumbled upon the hugely profitable concept of 'float'. In what started as just an incremental step towards solving a traveller's difficulties, American Express had become a fully-fledged financial services company.[42] It had created a truly new way to pay and benefited nicely from being the first to market.

These were the early days of a fundamental transformation of money, from silver and gold coins to paper – in other words, from literal to representational. Companies that participate actively in such a major transition period get far more than their fair share of benefit. And their resulting leadership position, at least in the case of American Express, is long-lasting.

In the 1940s, customers could not easily buy high-priced items without going to their banker and filling out complicated loan applications. You would walk into an appliance store, select a new refrigerator, then have to leave the store and go to see your banker. Along the way, you might notice another appliance store and end up buying your refrigerator there. You might even change your mind and not buy anything at all. The system was frustrating to the customer, as well as to the merchant (since there was a risk of losing your business) and even to the banker (who in many cases could have done more business if things were less complicated). The answer was the revolving credit card, which so successfully addressed the needs of the banker, the merchant and the customer that it launched an entire industry. At the time, a revolving credit card was a brand new way to pay, and substantially more convenient than the old ways. The first companies that launched credit cards benefited far more than their competitors, for a longer period of time. They are the origin of Visa and MasterCard.

The payment card industry has successfully initiated a number of revolutions since the 1940s, with the introduction of numerous innovations in payment card products. Common card acceptance brands, such as Visa, MasterCard, Diners Club and American Express, allowed customers to use their cards first across state boundaries and then across the globe. Once again, this was very definitely a new way to pay, better than the old way which was much more limited. During the 1970s, a magnetic strip was placed on the back of each card, containing an electronic copy of the information embossed on the card. Terminals placed at the merchant's checkout counter could automatically dial up a host to obtain a credit authorization, so the merchant no longer had to call an operator. This allowed banks to again offer their customers a new way to pay, more convenient and safer than the old manual way of getting a credit authorization.

THE ARRIVAL OF SMART CARDS

The smart card was invented over 20 years ago in France by Roland Moreno, an engineer exploring ways to automate identification procedures. The plastic credit card format was not his first choice. He tried using rings at one point but discovered that it was hard to precisely insert a knuckle in the reader and keep it still long enough for the transaction to take place. This is all part of the mythology that many people are familiar with, but what is the real story behind the success of smart cards in France?

New infrastructure technology requires the scrapping of existing equipment – cards, terminals, networks – and their replacement with new, expensive equipment. Whole industries must find justification to move forward or the budding technology dies quietly and disappears without anyone noticing. Getting the machine to begin inching forward, building momentum, is not something a single person can normally do. Of course Roland Moreno was not alone. In fact, the French government played a key role by acting as both supplier and buyer of the new technology. Government-owned chip and computer manufacturer Bull was encouraged to manufacture cards and terminals, while France's telephone company became a large customer for prepaid disposable telephone cards. This seeded the market and triggered the proliferation of French smart card and equipment companies.

At around the same time, French banks were faced with a dilemma. They were losing ground to large retailers who were launching their own credit cards and refusing to accept bank cards, as a way of cutting out the 2% to 3% processing fees. French banks were pressured to find a way to bring fees down to a level acceptable to large retailers. This became such a vital concern that all the banks came together as a single industry coalition, whose primary objective was to protect the French payment industry's overall market share. Smart cards were chosen as a technology that could dramatically reduce fraud and transaction processing costs, together representing the major causes of high fees. Pilots were launched and made to function correctly. Industry pricing was established across all players. Rules of competition were defined in order to cement long-term cooperation among banks and avoid creating significantly different card products that might fragment the payment industry and again weaken it. Of course, some of this could be illegal. The French banking community has built something that looks quite a lot like a cartel. In fact, in July 2004, the European Commission accused Groupement des Cartes Bancaires (CB), the operator of the French bank card system, and nine of its largest member banks of conspiring to restrict new entrants and stifle innovation in the French cards market by imposing complex fees and regulations that prejudice rival issuers. But the project has been very successful anyway.

In France, all Visa, MasterCard and Carte Bancaire debit cards have been equipped with microprocessor chips since 1992. Bank card fraud has decreased by over 80%. Transaction processing costs have also been reduced dramatically, since far fewer requests for authorization are required between the terminal and the central processing system. The workload on the central system is lighter, telecommunication costs are lower, and the checkout counter in the store runs faster and more smoothly.

It's amazing to me how quickly smart cards became totally accepted in French daily life over a very short period of time. I moved to France in 1989, just as chip cards were gearing up for national deployment. There was still a lot of talk at the time about whether or not the cards would really take off. 'There are too many problems,' people would say. 'My card was full the other day and I couldn't do any more transactions,' someone would add. 'And I keep forgetting my PIN code.' Or, 'There are too many terminals that kill the chip. They'll probably never get the terminals to work right.' These and many other problems were eventually solved and the trials were made to function properly. By the end of 1992, all French bank cards had been converted to chip.

I still keep a bank account in Boulder, Colorado, my home town, and use my Bank One Visa debit card in France. Sometime around 1996, I began noticing that merchants would occasionally become confused when I handed them my card. They might notice something different, maybe that it lacked a chip, but would go ahead and try to insert it in the chip reader anyway. Once it dawned on them that it was a foreign card, they would revert to the old card swipe wrist movement and complete the transaction. That was only four years after all French bank cards had chips. At first I found this a little curious but it didn't strike me more than that. It didn't happen very often. Perhaps these merchants were young, or new to handling cards.

By 1999, not a week went by that this did not happen to me. It even got worse. Cashiers would tell me that they wouldn't accept my card because it didn't have a chip. 'But I bought lots of things here last week and you were happy to use my card,' I would say. A manager would sometimes be called over. 'Yes, it's OK, it's a foreign card,' the manager would explain to the cashier. 'You have to use the magnetic strip.' Wait, there's more. Merchants in France have lost the card swipe wrist movement. They often don't swipe fast enough, or they swipe too fast, or they swipe the side of the card that doesn't have a magnetic strip. They rub the card on their sleeve then try swiping it again. I have become used to this. When someone looks confused I explain calmly that the card has to be swiped. When they can't get the wrist movement right, most merchants are happy to let me take the terminal and swipe my card through myself.

In just a handful of years, people had adapted so well to using chip cards that nobody thinks about them anymore. The best designed technology disappears, gets out of the way, and lets people get on with whatever they're doing. Getting French merchants to go back to swiping cards would prove to be very disruptive.

Not too long ago, people in the UK were saying that consumers would never adjust to entering PIN codes at the terminal. In December 2004, just one month prior to the liability shift, UK retailers were still worried about losing Christmas sales due to customers forgetting their PIN. UK banks are now convinced that customers will remember their PIN codes for maybe one card, and will avoid using multiple cards and remembering multiple codes (UK banks are preparing to allow customers to define their own PIN codes so they can avoid this problem). People in other countries have said that merchants and older people are going to have a tough time adjusting to the card being inserted chip side first rather than being swiped. When I first began developing marketing concepts around smart cards in France, people would say, 'Those are American marketing ideas that would never work in France, why don't you go back to the US and try selling them there?'. Several years later, presenting the mature software products in the US, some people would say, 'I'm not sure American merchants and bankers will catch on to all this, it might work fine in France but over here it's different.'

Today, smart cards are just one of the most recent of a long list of payment card innovations. Financial institutions expect smart cards to help reduce fraud and processing costs, fight the commoditization of payment, better compete against each other and take market share away from cash and cheques. The cost of handling cash is estimated at 2.6% of the purchase amount, for an annual cost of US$60 billion in the United States alone. Electronic cash is a major market that many financial institutions hope to play a part in, if not dominate. These are some of the reasons why Visa and MasterCard have been working together to define common EMV specifications.

Here is a description of the type of payment product that MasterCard is aiming at:[43]

> Ancient forms of value exchange – livestock, giant stones, spices, gold – were anything but convenient. Currency represented a fundamental shift away from these early varieties of money: You had something that represented the wealth you owned. The development of checks and bank drafts made carrying cumbersome amounts of currency less necessary. Today, credit and debit cards make life more convenient; they're easy to carry and use, and enable you to keep track of your spending by providing monthly statements. Virtual – or electronic – money takes the convenience concept a step further: You'll be able to access your bank account from anywhere in the world, obtain special services and benefits, and make purchases safely.
>
> In the future, convenience will drive the form of payment people choose. Credit cards let you buy now and pay later, if that's your preference. Debit cards let you buy now and pay immediately through your checking account. Electronic cash lets you 'pay in advance' by loading value onto the card, then making purchases using the stored value. 'Smart cards' will offer even more conveniences that cash and checks can't. A single card will store information – health insurance data, motor vehicle permits, currency exchange rates, loyalty programs – and operate as virtual money.

According to Edmund P. Jensen, President and CEO of Visa International,

> Cash and checks will become endangered species. Demand for convenience, globalization and large new markets are driving technological advances to create new ways to pay. Chip cards will transform the way we shop, allowing consumers to conduct safe electronic transactions anytime and anywhere in the world.

Visa is creating a new generation payment product based on smart cards. They call it a 'Relationship Card':[49]

> You carry a Banking/ATM card to access funds on deposit, a credit card for purchasing and cash for small value expenditures. A Relationship Card will do all that and more, enabling you to maintain a portable relationship with your bank, with access to all your accounts, at any time. In addition to multiple account access, the card will let you obtain account information, transfer funds between accounts and load currency onto the stored value component, eliminating the need to carry coins or exact change.
>
> Loyalty programs – like airline frequent flier, rental car VIP and video store frequent renter programs – reward customers for incremental card and/or product usage. Thanks to its microchip intelligence and macro capabilities, a Relationship Card will track and store data for loyalty programs automatically using transaction information as each purchase is made. Imagine having real-time updates on loyalty point balances and instant access to rewards as they become available ... 24 hours a day, all over the world.
>
> Just as you expect communications flexibility from telephone service providers, you expect banking flexibility from your financial institutions. You want access to customized financial products and services through electronic devices such as screen phones, personal computers, personal digital assistants, shared ATMs, point of sale terminals and other future devices. You want help in managing your various accounts, guaranteed security and control over your personal finances. Relationship Cards will deliver all this and more.
>
> The power and intelligence of the microchip will enable a Relationship Card to be more than just a financial tool. It will also be able to store a myriad of personal information that you might otherwise carry in your wallet. It could act as an all-in-one repository

for medical and dental insurance information. You decide what information you store, and a security system designed into the chip means data stored in the cardholder information part of the chip can only be accessed by you or those whom you designate.'

Since the payment card industry has already successfully integrated numerous innovations over several decades, much data is available to better understand how the move to smart cards is likely to succeed. It turns out that a surprisingly small number of intuitively obvious factors are required in order for a new financial card product to succeed. A new card product can substantially increase its chances of success by addressing the core motivations of each of the major participants. At the risk of oversimplifying: banks want to increase their share of transactions, merchants want to increase their share of profitable customers and customers want access to high-quality products and services at the lowest price. The most successful products satisfy all three requirements. Unsuccessful products satisfy a single party's requirements, perhaps two.

EXPANDING THE PAYMENT CARD MARKET BY ADDRESSING THE NEEDS OF MERCHANTS

In the 1980s, first in the US and then over much of western Europe, new retail locations became increasingly hard to come by. For decades prior to that, one of the easiest and most effective ways for a retailer to increase sales and profits had been to simply open new stores in new locations. Once that became less of an option, retailers had to find new ways to increase revenue. They had to begin focusing on growing same-store sales. One solution was to give customers an explicit incentive to be loyal to the retailer's store as opposed to the competitor's store.

A well-defined and well-managed loyalty programme, launched before competitors get into the game, often proves to be a useful weapon that can deliver significant results. In just a few years, UK supermarket retailer Tesco was able to dethrone Sainsbury's, a company with a long and famous history that was the undisputed market leader at the beginning of the 1990s. Tesco proved adept at introducing a range of innovative new shopping services to entice customers. In 1995, it launched an effective marketing campaign centred on the Clubcard, Tesco's loyalty card. Two years later, Tesco formed a joint venture with Royal Bank of Scotland and launched Tesco Visa and MasterCard credit cards which let customers collect Clubcard points on all purchases. By 2003, the Tesco joint venture with Royal Bank of Scotland was managing 4.6 million credit card customer accounts in the UK.

When Tesco first launched its Clubcard, Lord Sainsbury famously dismissed the idea as nothing more than an 'electronic version of Green Shield stamps'. He lived to regret that comment. Sainsbury's was forced to make an embarrassing turnaround and introduce its own loyalty card when the Clubcard became a roaring success. But it was already too late. Sainsbury's was forced to play catch-up, with decidedly mixed results. Sainsbury's Nectar card, launched only in 2002 in partnership with Barclaycard, was far too slow in coming.

In October 2004, Sainsbury's announced lay-offs at its head office, the halving of dividends, store closures and the suspension of new store roll-outs, and its first pretax loss in its 135-year history.[45] During the same week, Tesco announced that its half-year profits had surged 28% to a record £805 million. According to Sir Terry Leahy, the chief executive of Tesco, part of the reason behind the retailer's extraordinary success is its attention to 'intangibles' such as customer relationships.[46]

Shoppers have always enjoyed payment incentives, even at the birth of credit cards, when large retailers like Sears and J.C. Penney's began offering charge cards to their customers, and enticing shoppers to use them frequently in exchange for special discounts. Then, in the 1980s, US banks and financial institutions began tying their credit cards to airline incentives. The first co-branded programmes let cardholders earn travel incentives on one airline only. Marine Midland was first in 1986, followed by Citibank/American Airlines and First Chicago/United Airlines in 1987. Multi-participant programmes let cardholders earn travel on multiple, selected airlines. Diners Club was first to launch such a programme when it introduced 'Club Rewards' in 1984. American Express followed with 'Membership Rewards' in 1991, and Bank of Montreal and Citibank launched similar cards in 1992.

In fact, payment incentives go back even further. Much further. Merchants have always provided incentives to encourage customers to pay more, remembering each shopper's purchase history and preferences and offering customized incentives taking that personalized knowledge into account. The incentives have always been linked to the payment process, and they have virtually always been based on some fairly simple volume purchasing economics that are quite pragmatic and straightforward. Buy more from me today and I'll give you a discount. Buy a big box of detergent and the cost per ounce is less than what you would pay for a little box; a one pound can of coffee generally costs less per ounce than a small jar of coffee; 100 PCs purchased at the same time, or over a one-year period, cost less per PC than if you only bought ten. This is because there are fixed costs related to each transaction, regardless of the size of the transaction. High marketing costs are necessary to find a new customer, bring him into the store, convince him that the products and services are satisfactory and process the first order. Once that's done, all subsequent orders enjoy lower transaction costs and profitability is generated by reduced servicing costs, less price sensitivity, increased spending and favourable recommendations passed on to other potential buyers. On a per unit basis, volume transactions have lower fixed costs than smaller transactions.

All successful loyalty programmes are based on these same simple economics. But along the way, things got complex – much more complex. When stores became large international chains and shopkeepers became checkout clerks, there was nobody at the moment of payment who could remember an individual shopper's payment history and preferences as in the past. The payment process became industrialized and impersonal, mechanical and identical for every single customer. Still, lots of data was available – entire databases' worth – except that it came after the payment event had already happened. In order to process all the data, complex concepts have been developed by database marketing firms, building on top of the basic foundations of volume pricing. Lifetime value, for example, is the predicted future revenue expected to be generated by a customer based on that customer's current behaviour. To calculate lifetime value, a particular customer's prior purchases are summarized according to recency-frequency-monetary (RFM) parameters indicating how recently the customer last made a purchase, how often purchases are being made, and what cumulative amount was spent over a given period. Lifetime value and RFM are certainly important concepts, but they are really nothing more than complicated ways of measuring volume purchasing. Loyalty marketing may be shrouded in a lot of jargon, but the basic economics surrounding loyalty are no more complicated than those surrounding the simplest volume pricing policies. These concepts have become complicated because of the necessity to do much of the processing after the payment event has taken place. Direct marketing professionals use lifetime value and RFM because they need customer segmentation methods to target mailings. Using centralized databases and sophisticated data mining software, customers are ranked according to how recently they last shopped, how many purchases they made over a given period and how much they spent. Coupons, gift certificates and rewards can then be sent to customers based on individual or group behaviour profiles. The method is well established and relatively effective for higher-value purchases capable of justifying the high cost of operating the service.

Airline frequent flier programmes all use these techniques. They are all modelled on precisely the same volume purchasing economics that exploit diminishing fixed costs per transaction. Repeat customers are less expensive because acquisition costs are spread out over more transactions. Repeat customers also tend to buy more seats at full price and higher-margin products.

Volume purchasing in virtually all industries and markets displays the same characteristics. Higher-volume purchases cost proportionally less in fixed overhead and represent a proportionally larger share of profits. Loyalty is nothing more mysterious than this. It is simply a way for customers to enter into a business agreement with a supplier who essentially says, 'If you spend more over time with me and buy more products more often, you pay less.' In business-to-business relationships, a vendor that does not offer volume pricing would be considered strange. In business-to-business this is not called customer loyalty; it is just standard business procedure.

Many US supermarkets have two prices next to products: the normal price, and the special price you pay if you spend at least $20 today. This is because analysis of the details of each transaction shows that larger shopping baskets represent a proportionally higher share of margins. Discount supermarkets were originally designed for customers who shop on average once a week and whose transactions were naturally larger, usually well above the $20 advertised today. As more and more supermarkets were built, bringing them closer and closer to people's homes, and more and more began staying open seven days a week and even 24 hours a day, customers began treating supermarkets as convenience stores. Run down to the store for a quart of milk. Stop by the supermarket on the way home from work for a pack of beer and maybe a movie. When customers shop at traditional convenience stores like 7 Eleven, they are accustomed to paying significantly higher prices than in traditional supermarkets. This has prompted many grocers to say, 'If customers appreciate our stores as convenience stores, they should be willing to pay convenience store prices.'

When loyal customers are well rewarded, unanticipated benefits occur. For example, in 1980, just before airline frequent flier programmes became ubiquitous, passengers would tolerate an average delay of approximately 30 minutes before switching airlines. In 1990, after most passengers had joined at least one frequent flier programme, the tolerance to delay shot up to an average of three and a half hours, an increase of seven times.

Offering air miles and other enticements in exchange for increased payment card usage has generally proven to be an effective way of differentiating payment products and competing on benefits rather than interest rates and transaction fees. Banks have been very successful at convincing large retailers to outsource their payment processing activities for private label and co-branded credit cards. In 1996, 61% of US store card receivables were processed in-house by retailers themselves. Retailers increasingly shifted their strategy to focus on their core business, leaving credit processing activities to third-party financial institutions. By 2003, only 16% of store card receivables were processed in-house.[47] By that time, there were about 75 million co-branded cards in the US – 12% of the total 650 million MasterCard/Visa cards, but accounting for 25% of all expenditure on bank credit cards. Customers spend almost double on co-branded cards than standard credit cards. Average monthly spending on US co-branded cards was US$452 against US$239 on standard cards. In many regions, the main driver of recent growth has been private label conversion, by which retailers rebrand their in-house private label programmes with MasterCard or Visa.[48] Large retailers capable of launching cards with their brand on the front, either private label or co-branded, have been the source of substantial credit card growth over the past 50 years.

Once banks began offering airline rewards, the whole industry quickly saw the benefits of such programmes and companies scrambled to establish alliances with major airlines. Opportunities for co-brand relationships in all retail sectors, not just airlines, are obviously limited. Within ten years after the launch of the first co-branded airline card, there were virtually no more exclusive agreements available anywhere in the world. Banks and financial institutions had completely secured the market. Similar market saturation has since occurred across most categories of retailers capable of justifying issuing their own credit cards. However, there is a much larger merchant category that has not yet benefited from private label or co-branded cards. This group consists of retailers who are unable to justify issuing their own credit cards, primarily because their spending category does not represent a sufficient portion of an average household's monthly expenditures, making such cards less interesting to customers than general-purpose payment cards. In terms of number of retail outlets, this group is by far the largest and tends to still be dominated by cash and cheques. For banks, this new market could be even bigger than private and co-branded cards.

The vast majority of merchants in this category still cannot efficiently identify and reward their customers. Payment incentives for low-value transactions, say transactions under $20, cannot be managed using the expensive methods of direct marketing professionals. A major difficulty for card issuers is that, for this category of retailers, analysing reams of customer behaviour data for small transaction values and applying complicated direct marketing techniques is too cumbersome and costly. Many marketing techniques were developed for private and co-branded cards, all substantially the same, all based on points, miles or cash back calculated centrally, with monthly statements sent to customers informing them of their benefits, and with complicated marketing concepts used to analyse customer transaction data for targeted mailings. But instead of blindly copying these expensive and cumbersome techniques, all developed for other retail categories which enjoy bigger transaction sizes, in such cases as airlines, much bigger transactions, there is an easier solution. Rather than copying direct marketing techniques used by frequent flyer co-brand programmes, rather than trying to apply direct marketing techniques to low-purchase value merchant segments ... why not simply go back to the underlying concepts which were the foundation for all payment incentive programmes?

Most of these merchants already resort to paper promotions, such as coupons for discounts, or cards punched or stamped at each visit, offering a free item or discount after a certain number of visits or dollars spent. Like airline travel incentives, paper cards punched at each visit are based on the same economics that reward volume purchasing by sharing those savings with the customer. Unlike incentives managed through database marketing techniques, they are much cheaper to operate because they require no sophisticated direct marketing support. How many frequent-customer punch cards do you have in your wallet right now? Your favourite sandwich shop might have given you one that offers a free sandwich after buying several. You may also have a card from your local coffee shop, your dry cleaners and perhaps the video rental store. Cinemas give you one for free popcorn or a drink. Fast food restaurants use the technique to offer a free meal. Parking lots offer free parking after paying for a certain number of days. Using paper cards to manually count up visits or purchases, rather than doing so electronically with a plastic card, is not the only thing these merchants have in common. In many cases, they are primarily cash and cheque merchants.

Private store cards and co-branded cards tend to be used by large, centralized merchants who are able to justify wallet presence because they enjoy higher transaction amounts, while paper loyalty cards tend to be used by merchants with smaller transaction amounts, who don't necessarily accept credit cards and who are generally unable to get customers to sign up for their own private or co-branded card and keep the card in their wallets. Financial institutions have been exploring how smart cards can help increase the use of credit and debit cards among these merchants by reducing the cost of processing low-value transactions and making it feasible to once again offer personalized payment incentives linked to the moment of purchase. An enhanced payment transaction performed offline, using full-featured EMV cards, removes much of the complexity that has been required to process payment incentives in the past. With EMV cards and real-time payment incentives, there is no need to analyse reams of transaction data. There is no need for complicated and costly direct marketing campaigns. The payment process becomes rich once again with store personnel capable of providing personalized payment incentives on the spot, at the moment of purchase, taking into account each customer's individual purchase history and preferences. EMV used in this manner actually makes the whole process of offering payment incentives simple once again, because all of the necessary information is available at the moment of purchase. If a customer is entitled to a free product because this is their fourth purchase in the same month, then the product is given to the customer on the spot – it's that simple. In the same way that private store cards and co-branded cards successfully addressed the needs of major retailers by linking payment with incentives like airline miles, new generation credit or debit cards with chips for offline transactions and for managing payment incentives will prove to be a powerful combination to break open the much larger market of merchants that predominantly accept cash and cheques.

Fast food restaurants, coffee shops and dry cleaners are not the only companies to use paper cards that are punched or stamped at each visit. For decades consumer goods manufacturers have dreamt of techniques that would allow them to give discounts to customers after a certain number of purchases. Several generations of Americans have cut out proof-of-purchase seals from cereal boxes, coffee labels, detergent and countless other products. They paste them in a booklet and send them back to the manufacturer for a coupon for a free item. The technique is complicated, messy and very expensive. Often, the same core group of fringe consumers participate over and over again in these programmes. Smart cards can help bring these techniques into the mainstream. Why not allow Procter & Gamble to run their own electronic incentive, exactly like a sandwich shop's programme? By running the programme on the customer's smart payment card, administration becomes very simple. The customer pays with their credit or debit card, and the receipt shows an additional line of information that tells the customer how many more dollars they need to spend with Procter & Gamble in order to receive their gift certificate or electronic coupon.

A real need exists for this type of programme. Consumer goods manufacturers in the US spend over US$7 billion every year on coupons. Consumer goods manufacturers and retailers both detest cents-off coupons. Manufacturers consider coupons an inefficient way to spend their marketing money. Redemption rates of one to two percent of coupons distributed in newspapers is a great example of the inefficiencies of the system. Many attempts have been made to improve the process. In 1985, Citicorp launched an ambitious project to put coupon cards in the wallets of 40 million US consumers. By means of supermarket scanners, linked to a central management system, they hoped to build up a wealth of information regarding the buying habits of cardholders. Consumer goods manufacturers would be able to offer discount coupons to buyers of competing products, or to customers who buy a certain quantity of a given product over a period of time. Five years and US$200 million later, the programme was only able to attract two million cardholders, 5% of its target. The centralized approach turned out to be too heavy to manage. Customers received confirmation of acquired benefits long after they had made a purchase, so showing the card at the cash register had no direct immediate link with obtaining the discount.

In 1993, the world's first e-coupon smart card, called the 'PromoCarte', was tested in French hypermarkets. The customer presented their card to the cashier, who inserted it in the same payment terminal that handled bank cards. Scanned purchases were automatically compared to the coupons stored in the card's memory chip. When a couponed item was purchased, the system deducted the discount from the customer's receipt. Then, new coupons were added to the card, based on an analysis of scanned items, and a second customer receipt was printed, listing all the coupons available on the customer's card. Thirty consumer goods manufacturers participated, representing major brands in 50 different product categories – Nestlé, Procter & Gamble, Quaker, Heineken and Unilever, among many others. Redemption rates averaged 20%, almost ten times that of coupons printed in newspapers. Customers quickly learned to keep their coupon list for their next visit. Many even began writing their shopping list on the back of the receipt. This was completely unanticipated. Just a few short years after smart cards had become a common part of French life, consumers clearly understood that information like coupons could be stored in their cards, updated and removed automatically through a point of sale terminal. However, several drawbacks proved to be fatal. Customers neglected to present their card when they were not purchasing a couponed item. Every time a customer did not present their card, an opportunity was missed to issue new coupons. And since new coupons were not added to the card, the customer had even less of a reason to pull out the card at the next visit. Customers did not have enough coupons to choose from. We estimated that the number of available product categories needed to be doubled or tripled, to 100 or 150. The trial also showed that customers did not like having to pay with two cards, one from their bank and one for the coupon discounts.

In 1996, Procter & Gamble tried a different approach, simpler and theoretically more attractive for customers, merchants and consumer goods manufacturers. The company began an 18-month test in upstate New York, eliminating coupons in favour of everyday low pricing, nicknamed EDLP. The idea was that consumers would prefer consistently low prices everywhere and not have to chase coupons. Many packaged goods marketers watched with interest, and some also began scaling back their coupon activity as well. But it didn't work. Consumer demand forced Procter & Gamble to call off the campaign four months ahead of schedule. Coupons are a century-old tradition. Shoppers love coupons. They feel they are beating the system. This proved to be an insurmountable barrier for even industry giant Procter & Gamble. Some retailers point out that although many customers want coupons to continue, there are an equal number of customers who really see them as a hassle. They don't have the time to clip them, file them and manage them. Yet they feel guilty that they're not taking advantage of discounts that they are entitled to.

Coupons are processed as a form of micropayment at the point of sale. They are presented to the cashier during the payment process and are assimilated as a payment media, just like cash or cheques. US coupon-users report an average 11.5% savings on their grocery bills with coupons. Eventually, it would make sense for new generation credit and debit cards to extend their payment capabilities to also cover coupons. The key is not to simply make coupons paperless, but to make e-coupons much more powerful than the paper version, primarily by adding frequency and monetary value parameters that let consumer goods manufacturers give customers a bigger discount for spending more over time. During the PromoCarte trials, consumer goods manufacturers provided hints at the type of product they would be willing to invest heavily in. One manufacturer after another kept telling us that electronic coupons are a neat way of getting rid of paper coupons, but what would be really great is some way of building loyalty, just like retailers do with their frequent buyer programmes. What they were asking for in fact was their own electronic incentive programme. A customer might participate in several programmes with major consumer goods manufacturers like Procter & Gamble, Nestlé or Kraft. After spending, say, $200 on P&G products, the customer might get a $20 gift certificate, electronically loaded on the card. Today, there exists virtually no loyalty method that brand manufacturers can use that is practical and cost-effective. So budgets will be allocated from a strategic perspective, as opposed to the tactical perspective that governs couponing budgets.

ONLY THE FITTEST PAYMENT METHODS SURVIVE

The most mature smart payment card market is still in France, where the banking industry had all cards converted to chip by 1992. Today, the French debit card system is the world's most efficient. Fraud is virtually nonexistent. Transaction processing costs are lower than most other markets, including the US where economies of scale are vastly greater. This is because a significant portion of transactions are handled offline and cleared and processed at night. The cards can be used anywhere, even at parking meters and to pay highway tolls for very small amounts, as low as one Euro. You can use your card at any McDonald's restaurant in France. Even tiny fruit and vegetable stores have payment terminals. Since most transactions are offline and the customer types in their PIN code, there are no receipts to sign and the whole process typically takes a few seconds.

The banking community can go very far in winning market share away from cash and cheques by making their current payment products more efficient. A chip debit or credit card will open sizeable new markets without the hassles and risks of creating a completely new paradigm like an e-purse card that has to be reloaded at a teller machine whenever it's empty. Examples in France of doing micropayments with a common debit card are striking.

The success of French banks prompted financial institutions in many countries to begin moving to smart cards. Live tests began outside of France around 1995. Initially, financial institutions were attracted to the electronic purse, or stored value systems that use bits in the card's microprocessor chip to represent cash. UK banks led by Natwest launched Mondex and created an international marketing phenomenon. MasterCard purchased 51% of Mondex, which had a snowballing effect on the number of pilots being carried out worldwide. Visa Cash has also been piloted in a number of different environments, one of the earliest being the 1996 Olympics in Atlanta where NationsBank, First Union and Wachovia issued thousands of smart cards. Attendees used the cards to buy soda, sandwiches and other small items. The most successful e-purse product by far was Proton, which was created by a Belgian bank consortium. In 1999 Proton had become the world's foremost e-purse with 30 million cards distributed by over 250 banks, and accepted by 200 000 terminals in 15 countries. Proton was one of the few e-purse systems to have been rolled out nationwide, with full deployments in Belgium, Holland, Switzerland and Sweden.

Table 5.1 compares two major smart payment card programmes, France's Carte Bancaire debit system and Belgium's Proton e-purse. Statistics show acceptance levels two years after nationwide roll-out.

Table 5.1 Comparison of smart debit card and e-purse

	Smart debit card	**e-purse**
Card programme	Carte Bancaire	Proton
National test market	France	Belgium
Population	50 million	10 million
Cards issued 2 years after rollout	21 million (42% of population)	4 million (40% of population)
Merchant terminals installed	520 000	23 000
Terminal penetration ratio	1 terminal for every 96 inhabitants	1 terminal for every 435 inhabitants
Total purchase transactions during the first 2 years after launch	2.5 billion (119 per cardholder)	30 million (7.5 per cardholder)
Transactions per month	130 million (6.2 per cardholder)	2.3 million (0.6 per cardholder)

Comparing the two, the French debit card system enjoys a healthier level of acceptance among merchants and cardholders. Over the first two years after launch, French debit cards generated 16 times as many transactions per cardholder as Belgian e-purse cards during Proton's comparable post-launch period.[49] There were proportionally 4.5 times more places to use the card (one terminal for every 96 inhabitants versus one terminal for every 435 inhabitants). The French system also works better because it does not represent a paradigm shift. It is an evolutionary improvement that uses new technology to leverage the financial industry's core payment-processing competencies.

General-purpose electronic purse systems like Visa Cash, Mondex and Proton are built to closely resemble cash. Money is withdrawn from an automatic teller machine-like device and stored in the card's chip. Each time the card is used for payment, money is taken out of the chip and electronically stored in the merchant's payment terminal. Once the money is used up, the card must be reloaded again. E-purse is truly presented as a straight substitute for cash, a new form of money. Mondex goes the furthest in this direction by making transactions completely anonymous, just like cash, keeping no trace of where and how a card is used for payment. General-purpose e-purse cards have not achieved sufficient market acceptance precisely because they are merely a cash substitute, with unclear additional benefits over pocket change. The benefits are certainly there, but they are very difficult and complicated for merchants to buy into. Since it is a change in the form of money and not in its substance, we can expect to see the e-purse have trouble in the mainstream but possibly find some significant success in vertical markets at the fringes of mainstream commerce. This is precisely what we see with the use of the electronic purse in closed environments like launderettes, cinemas, campuses and payphones.

Although e-purse as a bank card product has had a hard time justifying its existence, e-purse used in a closed environment has proven to be somewhat more useful. Prepaid telephone cards in many countries have achieved very respectable adoption rates. Transit authorities use contactless smart cards to speed commuters through trains and buses. French cinemas have been issuing reloadable e-purse cards since 1987, to the tune of 100 000 cards per year. Club Med and countless other companies, universities and college campuses, use them as a way to make paying in a closed environment easier for customers.

How does a closed environment e-purse fit in with a bank's strategy? More often than not, it doesn't. The core issue is to determine who manages the float. Telephone companies have become used to holding on to the customer's money for quite long periods before the card is actually used up. In France, an estimated 18% of prepaid amounts are never used! The amount sits as residue on the card for an indefinite period of time. Cinemas have also grown comfortable with the practice of managing the float on their cards. Financial institutions will have great difficulty acquiring and managing the float generated by a closed environment e-purse. During the mid-1990s, several e-purse projects linking telephone companies or transit authorities with financial institutions were under preparation, in the US and several other countries. Most of these discussions never resulted in a tangible implementation. The main problem concerned float.

'I'm a financial institution,' says the bank. 'It's my role to manage the float, plus my business case depends on it. Why else would I issue e-purse cards?'

'Perhaps,' says the transit authority. 'But those are my trains, buses and customers. I will manage the float.'

Statistics comparing smart debit cards to e-purse schemes show clearly how much easier it is to get merchants and customers to adopt a new payment card product if it does not require too great a change in behaviour and if it provides very clear benefits that all parties recognize easily. For most merchants, the e-purse benefit of reducing the costs of handling change is a concept that appears quite theoretical. As for customers, honestly, how big a problem is change in your pocket? Enough to endure yet another paradigm shift?

A major electronic cash pilot in Manhattan ended in 1998 after one year of tests jointly run by Citibank, Chase Manhattan Bank, Visa and MasterCard. Sponsors of the trial say they learned valuable lessons about technology and usage patterns, even though the trial wasn't a huge hit with customers or merchants. Citibank and Chase issued a total of 100 000 cards loaded with electronic cash. More than 600 merchants were recruited to accept the cards for payment. After 14 months, most consumers never reloaded their cards with electronic cash and two-thirds of the merchants ended up abandoning the trial. A number of people have observed that the trial was hampered because residents of the Upper West Side, where the trial took place, couldn't use their smart cards when they went to work in other parts of Manhattan.

The trial was hamstrung by the lack of an attractive value proposition. After the trial, many bankers and smart card experts came to the conclusion that consumers need a financial incentive to use the cards, such as those provided by loyalty or reward programmes. Dudley Nigg, executive vice president of Wells Fargo Bank, speaking at CardTech/SecurTech 1998, said, 'In Visa Cash and Mondex, we still don't have a really viable loyalty proposition – and that's one reason why smart card pilots haven't been successful.'

A general point of view following customer and merchant reactions to numerous e-purse trials in the United States and elsewhere is that people don't need an e-purse. It's pretty easy to write a cheque, hand over cash, or pay with a credit card. Harvey Rosenblum, senior vice president and research director at the Federal Reserve Bank of Dallas, says that 'the existing payment mechanisms that people know and trust work very efficiently and are fairly cheap'. Rosenblum thinks smart cards will appeal to individual customers when they offer perks and rewards to frequent customers.[50]

The e-purse pilots launched by banks in the late 1990s are generally considered to have been too costly for the benefits they provided. The benefits for consumers and merchants were never strong enough to create healthy demand. Getting customers to load money to their card has proven to be very difficult. The vision of generating interest on the money that consumers load to their cards was a powerful magnet, but has not benefited banks.

Soon after the plug was pulled on the Manhattan e-purse project, at the height of the Internet e-commerce bubble, a number of financial institutions began using smart cards to help with shopping on the Internet. High-profile products were launched, such as the American Express Blue card, the Providian Smart Visa card and the Smart Visa card issued by US department store chain Target. The plan was for customers to use smart card readers attached to their PCs for more secure web purchases, storing their favourite websites and downloading coupons and advertisements from the web. Later, when the customer showed up at the store, the coupons stored in their card would be automatically applied to their purchases.

This was during the heyday of the Internet bubble when retailers were projecting astronomical growth rates for online shopping. Although there was a definite slowdown when the bubble burst, online shopping continued to grow. According to Jupiter Research, US online retail sales are expected to grow from US$55 billion in 2003 to US$117 billion in 2008.[51] This would still be a drop in the bucket in terms of overall retail sales, barely 5% by 2008. But the influence of the web is greater than that. Consumers continue to research purchases online before buying in store. Jupiter Research's forecast projects that, by 2008, nearly 30% of offline retail purchases will be influenced by research performed online. It could make sense to help customers select products and related coupons online before going to a Target store to shop.

Target spent two years and an estimated US$40 million on their smart card-based coupon system. They retrofitted 37 000 cash registers to handle nine million new cards and installed coupon kiosks at 1191 stores. But, for average shoppers, the technology was not simple and straightforward to use. Shoppers had to order a special card reader from Target, then install the device on their home computers and use it to download coupons to the cards. Or they could obtain discounts at in-store kiosks, which were unfortunately not very easy to find in the store. Neither harried parents nor techies gave the system more than limited use. It was easier for customers to look through the Sunday paper and clip the coupons they wanted out of a selection that was far greater than anything Target could put on their kiosks. In March 2003, less than a year after the programme was fully up and running, Target announced that the cards' smart functionality had experienced limited use and the retailer would return to traditional magnetic strip cards.

Target was requiring customers to make too big of a change in how they did things, similar to how the Manhattan e-purse project required customers to reload their cards, which most customers never did. There are other similarities. In the same way that the Manhattan test did not provide customers with enough places to use their cards to make reloading them worth their effort, Target's deployment did not offer customers sufficiently exciting offers on a large enough selection of products to make downloading coupons worth their effort.

Randy Vanderhoof, Executive Director of the Smart Card Alliance, tried the card out like most other people in the financial services industry. Here is what he had to say:

> As one of those consumers who signed up for a card and reader and eagerly loaded the software and started surfing the Target.com site for

> exciting rewards, I must confess I was not enamored by the benefits Target offered to me as a consumer. A 10 cent coupon off my next purchase of a new toothbrush or $1 off a family-size box of Tide detergent did not get my consumer juices pumping. But the technology did its job. The software loaded easily; the reader worked the first time I plugged it in; and my smart card came to life on my PC for the first time in my career as an ordinary consumer. I am not a department store shopper so I never tried to use my card in the neighborhood Target store (I get my toothbrushes and detergent at the grocery store, thank you), so you can blame me in part for Target's decision. However, how about giving credit to Target and Visa USA for putting their reputations and resources on the line to try to make this work? Pioneers rarely receive the recognition they deserve, only the critics' labels that unfairly judge their achievements as failures.[52]

Amen.

Every smart card pilot has been valuable for the industry because each pilot has revealed a wealth of information on what customers and merchants really want in a new generation card product. Customers want something simple, convenient, accepted everywhere, with clear benefits, easy to use in many different outlets and very easy to understand. Merchants especially want to know that the new payment method they are being asked to accept will allow them to increase their revenues and their profits without creating hassles and undue complications. New card products that don't adequately address these requirements can't build enough steam to achieve critical mass.

Today, Visa, MasterCard and American Express are all encouraging banks to leverage their current credit and debit cards to win market share from cash and cheques, by upgrading them to low-cost smart cards which make transactions more efficient and offer real-time incentives that encourage customers to use the cards more often. This is powerful, inexpensive, simple to use and easy to operate.

MONEY IS POISED TO BECOME INTELLIGENT

In the past, money evolved to become more and more symbolic. Paper money and cheques are symbols void of intrinsic value (the paper costs nothing). Virtual money takes the same concept further by using electronic bits as the symbol for value. All forms of symbolic money are backed up by governments, legal bodies and financial institutions.

Smart cards introduce a brand new element. The bits are of course a symbol for money, that's not new. But now money in the form of a smart card payment instrument is able to store and understand information concerning itself, how it has been used in the past, where it has been used, when, and how often. Suddenly, money becomes intelligent by its ability to watch and monitor itself and its user, reacting to stimuli that it and its user have generated. It can cause things to function differently if the card has been used frequently over a period of time, or if it is the first time it is used at a particular place. It can trigger warnings, and reminders, either to its user or even to itself.

Many readers will recognize this as a classic feedback loop, the process of using the output of a device as new input for the same device. Feedback loops are at the heart of many of mankind's most useful and revolutionary technologies. James Watt created a regulator for steam engines that used a feedback loop to stabilize the motor at a constant speed of the operator's choice. This launched the Industrial Revolution. Telephone engineer H. S. Black, working at Bell Laboratories, created the first electrical feedback loop in 1929 in his search to improve amplifier relays for long-distance phone lines. This paved the way for the invention of vacuum tubes and transistors, the building blocks of the information revolution.

Feedback loops appear naturally in biological processes. Neurons, DNA cells and higher organisms all depend heavily on feedback loops. In fact, not a single example has been discovered of a biological process devoid of feedback. It also appears in complex social structures and in economic theory.

Feedback loops allow an entity to become self-governing, self-adjusting and capable of adapting to many different and unpredicted stimuli. When layers of feedback loops are piled on each other, resulting in the complex organisms encountered in life, they allow entities to learn and adapt to the world in a way that we intuitively consider to be alive. Consciousness itself emerges from layers upon layers of feedback among several billion neurons that recognize nothing more than two states, charged and discharged. When feedback is applied to very large numbers of ridiculously simple entities, the overall system becomes excruciatingly complex. When systems become sufficiently complex and interconnected, they self-assemble into a new, higher order, which is more than the sum of its parts. Human engineered entities are just beginning to become complex in a similar fashion, as they integrate layers upon layers of feedback.

Once they were discovered, feedback loops were introduced in more and more human engineered objects, resulting almost always in spectacular innovations. Taking the long view, using smart cards to create a feedback loop for money was inevitable.

As the substance of money is transformed and becomes increasingly intelligent, it will adapt to provide its user greater services and convenience. Throughout humankind's relationship with technology, whenever feedback loops were built into human engineered objects, floods of innovations appeared that were completely unanticipated. This is about to happen with money.

INTELLIGENT MONEY: A GLIMPSE OF THE FUTURE – FOUR KEY OBSERVATIONS

1. Money has already undergone three stages of fundamental transformation, from literal, to representational, to virtual. Smart cards are poised to trigger the fourth transformation. Money is about to become intelligent. Companies that participate actively in such a major transition tend to secure a long-term competitive advantage in their industry.

2. Merchants have always provided incentives to encourage customers to buy more, remembering each shopper's purchase history and preferences and offering customized incentives. The incentives have always been linked to the payment process, and have virtually always been based on simple volume purchasing economics that are pragmatic and straightforward. Buy more from me today, and you will pay less per unit.

3. In the same way that private store cards and co-branded cards addressed the needs of major retailers by linking payment with incentives like airline miles, new generation cards with chips for offline transactions and for managing payment incentives will prove to be a powerful combination to break open the much larger market of merchants that predominantly accept cash and cheques. Full-featured EMV cards remove much of the cost and complexity that has been required to process payment incentives in the past. The process becomes rich once again with store personnel capable of providing personalized incentives on the spot, at the moment of purchase.

4. Money in the form of a smart payment instrument is able to store and understand information concerning itself, how it has been used in the past, where it has been used, when, and how often. Money becomes intelligent by its ability to watch and monitor itself. This is a classic feedback loop of the type that has produced many of mankind's most useful technologies.

CHAPTER 6

Key Factors for a Profitable EMV Deployment

Flowing water avoids the high ground and seeks the low ground … avoid difficult methods and seek easy ones. Do simple things well and quickly.

Sun Tzu

Is an EMV deployment successful when all of the technical problems have been eliminated? When fraud is under control, cut in half, or cut by 80% or some other factor? Is an EMV deployment considered a success when the budget is respected and the project is completed on time?

In this chapter, the definition of success will be more ambitious than all of these measurements. Certainly, all of the technical problems must be eliminated and fraud must be dramatically reduced. Any EMV deployment which does not succeed in getting fraud under control cannot be considered a success. The deployment budget must be respected and the project must be completed on time in order to qualify as a success.

But the move to EMV is such an important development that we as an industry need to be more demanding than that. If a deployment is successful according to all of the prior criteria, but fails at creating differentiation, fails at positioning payment as an integral and strategic tool that justifies merchants continuing to pay ongoing transaction fees and commissions, fails at getting merchants to look at payment as anything but an intolerable expense that needs to be cut, then the deployment cannot truly be considered a success.

Learning from the experience banks are having with EMV, it is apparent that a few key factors together can bring about a successful and profitable deployment. Of course, the technology alone does not guaranty success. How the technology is used is even more important.

1. *Build enhanced EMV features into your infrastructure before you launch.* This will lower your overall costs and maximize your return on investment. Adding enhanced EMV features after you have already issued cards, deployed terminals and upgraded your processing systems may be more expensive later.

2. *Attract cardholders by creating a simple and exciting value proposition.* Make your EMV card product easy to understand and beneficial for all players. Focus on features and services with mass appeal and which provide hassle-free benefits at every transaction.

3. *Provide numerous opportunities for customers to use your card.* If your cardholder base is not big enough to attract leading merchants, license your enhanced payment brand to other banks. Achieve critical mass as quickly as possible.

4. *Attract merchants* by showing how your payment network can provide them a simple way to target their promotions, discounts and coupons to a large audience. Boost transaction volume within your merchant network by encouraging cardholders to shop more at those merchants.

5 *Encourage merchants to do their own marketing.* You don't need to be involved in helping merchants define every coupon or marketing message printed on receipts. Leverage the creativity of thousands of merchant partners, at virtually no cost and effort to the bank.

6. *Show merchants how your new payment infrastructure can help address important problems* that are big enough to be discussed at the board level. If you can turn the discussion around and address quantifiable problems that keep the retailer awake at night – problems so big that the cost of your services is ridiculous in comparison – you will succeed in radically transforming your business.

BUILD ENHANCED EMV FEATURES INTO YOUR INFRASTRUCTURE BEFORE YOU LAUNCH

Judging from the number of articles written in the general press and the amount of broadcast time spent on the evening news, smart cards generate a great deal of interest in countries actively migrating to EMV. A Google search for news on 'chip and PIN' reveals over 100 articles published over a one-month period in the UK. Here are a few examples taken from widely distributed UK news media targeting the general public, some of it conflicting, as with any such subject under close scrutiny by journalists looking for a meaty angle to scoop.

> *Retailers still not prepared for Chip and PIN*
> Thousands of retailers are calling for more time to adopt Chip and PIN guidelines, despite an extensive campaign by the government to boost awareness, a new report claims.
>
> *The Register*, 29 October 2004

UK's chip-and-PIN rollout on schedule
Christmas will be first big trial for the technology, predicts analyst. The UK's chip-and-PIN rollout is on track to meet its December targets of issuing new cards to 36 million consumers and converting 636 000 tills, according to a report to be released next week by the Association for Payment Clearing Services. From January 1 2005, retailers that still accept customer's signatures instead of their PIN codes will be liable for any card fraud perpetrated by customers, rather than their banks.

Computing, 3 November 2004

Is chip and PIN safe? Am I alone in being worried about the new Chip and Pin system? The new system seems to move the fraud liability from the banks to the retailers. But in fact it moves it from both the banks and the retailer and puts it completely on the customer.

This Is Money, 4 November 2004

Chip and PIN cards take off
Three-quarters of all cardholders now have at least one chip and PIN card in their wallet, according to figures released yesterday.

The Scotsman, 9 November 2004

'Tis the season to watch your wallet
Christmas is the time when criminals step up their activity around cash machines as customers queue to withdraw cash before going on shopping sprees and attending festivities and celebrations. Criminal attacks on the machines nearly doubled over the past year, according to the Association of Payment Clearing Services (Apacs). Fraudulent withdrawals have risen by 85 per cent to £61m, making this the fastest-growing type of fraud in the UK. The widespread use of the new chip and pin cards, currently being rolled out, will make it impossible for card information to be copied.

The Daily Telegraph, 17 November 2004

These articles have two things in common. They are heavily slanted towards fraud, and they are not paid for by banks' advertising campaigns. Lots of media attention exists around chip migration, much of it at no cost to banks. If your competitors are primarily focused on EMV's ability to combat fraud, you have an excellent opportunity to stand out from the crowd with a message built around 'a new way to pay'. If you launch a full enhanced EMV product offering, while at the same time your competitors launch basic card products limited to PIN code verification but otherwise not very different from existing cards, you will definitely secure a larger portion of the marketing buzz, without a doubt. For the simple reason that the media has a compelling desire to talk extensively about chip cards and your message will be much more exciting than your competitors', you will attract attention from journalists looking to write about things that are out of the ordinary.

The window of opportunity only exists during EMV migration. Once migration is completed, the marketing buzz will move on to something new. If you launch with the same message as everyone else, then come back in a few months or a year with a new product offering using EMV's full capabilities, the opportunity for significant amounts of free publicity will be lost.

Another reason for building a full EMV system from the beginning is that the incremental infrastructure cost will be negligible. If a car mechanic has to take an engine apart to fix something, he might as well fix several problems at the same time, for virtually the same cost. When negotiating with card suppliers, card personalization service bureaus, terminal vendors and so on, it will be cheaper to have them commit from the beginning to supply you with full enhanced EMV products and services. If not, once the deal is negotiated it will be difficult to persuade them to deliver more for the same price.

Cards need to be formatted during the personalization process to include the appropriate data structures which will be updated later, at each transaction. Most cards available today can be formatted in such a manner, but if you don't ask for it, your card provider may leave this step out of the personalization process. You don't need to know which merchants will be participating, nor the types of offer which will be made. These details don't need to be present in the card when it is issued. They are added dynamically, later, but only if the appropriate data structures were planned for when the card was issued.

By including the enhanced payment data structures in your cards before they are issued, you can eliminate the need for multi-application card management systems that manage the dynamic addition of new applets to cards later, after the card has been issued and is in the customer's wallet. This capability is otherwise known as 'post-issuance download'. It is one of the costliest features of multi-application chip cards. The most successful EMV deployments have avoided such features and have concentrated on building enhanced payment capabilities into their cards *before* they are issued.

Terminals need to be deployed with an enhanced payment application already integrated and ready to run. Once the terminal is installed, you can turn on all kinds of enhanced payment feature later, simply by downloading new parameters to the terminal, in the same way as you currently update a merchant's receipt details or security parameters such as floor limits. With a full-featured EMV infrastructure, a bank does not need to predefine merchant partnerships, rewards rules, customer benefits or any other such things before deploying the infrastructure. These can be added to the network later, or removed, at will, by turning on or off the appropriate parameters within your terminal management system, but only if such a system was planned for when the terminal was deployed.

ATTRACT CARDHOLDERS WITH A SIMPLE AND EXCITING VALUE PROPOSITION

Make your EMV card product easy to understand and beneficial for everyone. Build a unique marketing message around the concept of 'a new way to pay', a richer, more memorable payment experience which customers and merchants immediately recognize as something much better than the old way to pay. Create a brand for the enhanced payment features which your customers will soon be experiencing whenever they use their cards to pay. Use the same enhanced payment acceptance brand across all of your card products, whether credit, debit, ATM, corporate cards, etc. This will make it easier later for merchants to know which cards offer these new payment features. If an enhanced payment acceptance brand is already available in your market, you might want to consider using it on your cards. This would give you fast access to a potentially large merchant network.

Focus on features and services with mass appeal and which provide value at every transaction, such as 'cash back and surprise gifts', or 'points and treats', or any other language which fits your market, and which indicates a multitude of benefits and perks which are very easy to get. Customers everywhere want less hassle and more value. That is why they switch cards so easily. When Australia's Reserve Bank slashed the fees banks could charge on Visa and MasterCard transactions, immediately causing banks to cut the value of their rewards programmes, customers switched over to American Express cards when those were made available to them, as the cards were not covered by the fee mandates and were able to offer bigger benefits.

If you are already offering a rewards catalogue, you might be able to cancel it and let customers redeem their points for goods anywhere your brand is accepted. Since operating a rewards catalogue is expensive, you will be able to give customers more points for their purchases. Also, you can get merchants to give additional bonus points on top of the points you finance, since each merchant will be able to offer tiered points and concentrate their promotional budget on customers who really do spend a lot with that merchant.

Show customers clearly how your new card product is easy to use and simplifies their lives. In comparison, show how the old way to pay (which is still the way your competitors do things) is riddled with administrative complexity. Customers get the feeling that procedures are complex because companies want to make it harder to acquire a reward. Here's a complaint many marketing professionals have heard: 'If everybody got the reward, the company would go out of business.' Or another one: 'They're just misleading customers. They get people to buy something for the gift advertised on the box, and then they make it impossible once you read the fine print.' Excellent marketing concepts often get bogged down in administrative details. Show how your payment card stands apart and doesn't require the customer to fill out special forms to redeem a reward from a catalogue. Show how your card lets the customer get surprise offers instantly, at the point of purchase, without doing anything differently, simply using your card to pay as usual.

Build on habits that customers are already familiar with. Customers have developed a smooth relationship with payment cards, automatic teller machines and payment terminals. Requiring them to learn new things like how to reload their electronic purse at a kiosk or at an automatic teller machine raises powerful roadblocks that you will have to overcome. Many of these are unnecessary.

A popular piece of folk wisdom tells us never to simultaneously switch jobs, towns and spouses. Change one thing at a time. Get adjusted. Then go on to the next. Customers are familiar with using their cards to pay at point of sale terminals and to withdraw cash at automatic teller machines. Merchants are used to punching a purchase amount on their terminal's keyboard and swiping the customer's card. That's the starting point. Early implementations should stick as much as possible to these basic habits. Systems must be simple to use and must not require significant changes in how merchants and customers currently perform payment card transactions. Additional keystrokes at the terminal should be avoided if possible, as well as new operations performed at kiosks or automatic teller machines. US retailer Target spent two years and US$40 million installing coupon kiosks, then had to discontinue the project one year later due to lack of shopper interest.

Once customers grow accustomed to inserting their card and entering their PIN code, and have grown used to the idea of having information dynamically updated and stored in the chip, the market will be ready to learn new habits such as using a kiosk or a PC to see what information is stored on the card. Then, once people are familiar with that, you might eventually be able to get a few early adopters to perform a completely new type of behaviour: adding things like coupons to their card or modifying personal parameters stored in the chip. Build on what people are already familiar and comfortable with. Don't rush people into completely new ways of doing things.

PROVIDE NUMEROUS OPPORTUNITIES FOR CUSTOMERS TO USE YOUR CARD

Cardholders want access to a large variety of merchants offering significant perks. If you are a sufficiently large issuer in your market you can create your own acceptance brand, put it on your cards, and establish an agreement with a large merchant-acquirer to attract merchants to your acceptance brand. If customers with chip cards carrying your acceptance brand represent a significant portion of a merchant's customer base and total revenue, it will be much easier to persuade the merchant to give bigger rewards, surprise gifts and other perks. If not, you may need to license your acceptance brand to other card issuers for faster critical mass. If an enhanced payment brand already exists in your market, be one of the first to join it. This is one of the most effective ways to maximize merchant and cardholder participation, grow your payment business, and take market share away from other payment methods as well as from competitors who have decided to launch EMV only for fraud-prevention purposes.

People who have spent over ten years watching, participating in and comparing smart card deployments would probably agree that achieving critical mass quickly is likely to be the single most important factor for success. Small pilots all suffer from the same drawback, lack of places for customers to use their card and not enough customers for merchants to see a difference in their total revenues. The Manhattan smart card pilot jointly run by Citibank, Chase, Visa and MasterCard ended in 1998 after only one year. Residents where the trial took place couldn't use their smart cards when they went to work in other parts of Manhattan. Most consumers never reloaded their cards with electronic cash and two-thirds of the merchants dropped out of the trial. Since customers still had to carry cash for most places where they shopped, the promised benefit of eliminating cash was never achieved.

Pilots are sometimes useful for demonstrating that the technology works, but they are utterly useless for measuring impact on customer behaviour. If you are about to do a pilot, what exactly are you testing? If you want to test the technology, it would be better and cheaper to test the technology quietly in a laboratory-type setting. If you want to test the value proposition and the impact on customers, honestly, you are wasting your time and money with a pilot. Worse, you are probably signalling your strategy to your competitors, giving them the ability to get to market before you by skipping the pilot phase. The value proposition is closely linked to the number of participants, in other words the number of cardholders and the number of merchants. By definition, a pilot consists of a small number of participants, therefore making it impossible to test the value proposition.

The easiest way to succeed is to move quickly to become the first to achieve critical mass around an enhanced payment brand which is adopted by card issuers representing a dominant combined market share that is so strong that merchants have to join the network. In the first phase, perception counts as much as reality. If you announce an alliance by the top three issuers, who agree to use the same enhanced payment brand and provide a clear timeline for deployment, you have probably pre-empted the competition already, before a single card has been issued.

ATTRACT MERCHANTS WITH A SIMPLE WAY TO TARGET PROMOTIONS TO A LARGE AUDIENCE

Differentiate your acquiring activities by offering merchants a branded enhanced payment service. Build a unique merchant marketing message around the concept of 'a new way to pay that lets you target your existing promotions to your best customers'. This positions the added value within the payment experience – in other words, in your world. It will keep you in the driver's seat and will help you keep a higher portion of the added value for your company and protect the margins on your fees and commissions. With 'A new way to pay', the retailer will look at your services more from a strategic perspective than in the past, making it easier for you to sell and negotiate your acquiring services without as much pressure on price. Let the merchant know that your payment services are related to the basic, no frills, old way to pay and come at a price that the merchant already knows well, but that the pricing structure is different for all the new services that let the merchant offer targeted promotions using payment data in the chip.

If you have a large merchant network and there is no enhanced payment brand in your market, create a brand like iPay yourself and systematically place a sticker at merchant outlets as you upgrade their terminals for EMV. Explain to merchants that customers with iPay cards will be able to pay with credit or debit, just like everyone else, or they can choose to pay with points and cash back that they have collected through their bank's loyalty programme. All of these different payment methods are the same as far as the merchant is concerned. You can also explain that other services will be available to the merchant as iPay cards become widely available.

License your brand to large issuers, whether part of your organization or not. If your organization is the top issuer and the top acquirer in the country, it will be easier for you to get substantial synergies out of your dominant position and create a greater gap with your competitors. If your bank is not a major issuer, it is very important for you to get your brand onto a large number of cards issued by other banks. This will give you a large audience for the additional services you can sell to merchants. Once you have a large card base available with your payment network brand, go back to your merchants and sell additional payment-processing services which leverage the brand. You won't need to reinstall software or upgrades. All you will have to do is set the parameters for the enhanced payment services that a merchant wants.

Merchants already offer lots of different promotions, coupons and discounts. They already know very well what types of offer work best under different situations. They spend lots of money printing these offers and distributing them through mass market methods. Merchants offer discount coupons in the local paper, by mail, or distributed in their stores. These coupons are almost never targeted, and they are expensive. A half page ad in a single paper, with four coupons, costs between US$11 000 and US$14 000 for a single Sunday printing. A small page in a co-op mailing package costs between US$400 and US$500 for a single mailing to a few thousand households. An acquirer does not need to be an expert in retail marketing to provide merchants with a way of targeting their promotions better. All you need to do is offer merchants several triggers that they can use to target promotions, such as the first visit to the store or chain, or the second, third, fourth or nth purchase in a given period, or after spending a certain amount of money. Then let them match up the actual offers to the triggers. Your fee structure might be tiered, based for example on the number of triggers a merchant uses. Perhaps your current acquiring fee structure might allow the merchant to use a single trigger at no extra cost, with additional fees charged for additional optional triggers. Charge a fee each time the merchant changes the promotional message triggered at the bottom of the receipt.

Merchants want things to run real smoothly, without having to do much in the way of pushing buttons and analysing reams of data. Surprisingly enough, most merchants are not attracted to the idea of knowing precisely who their customers are and what they buy. Only very large, centralized retailers have made serious attempts at data mining. Even then, data mining in retailing has been relatively timid. Merchants are pragmatic. Does the system help deliver special promotions to my best customers, so that I can concentrate valuable rewards on customers that really do spend a lot in my store? Does it run on its own without my having to become a database expert (and without my having to hire one)? Can it run automatically without having to train store personnel to do lots of new things?

Say I own a restaurant franchise that is part of a leading chain of restaurants. 'Buy one meal get the second free' coupons are very popular. I would like to give this offer to customers who come to my restaurant at least four times a month. It is very easy for me to calculate a return-on-investment. I don't need to know each customer's name, address and purchase history. I just need to know how many coupons I gave out and whether or not my restaurant's overall sales have increased.

Show merchants how they can use your enhanced payment services to maximize the customer's motivation to make the next purchase. Merchants can increase the value of rewards as a customer spends more or shops more frequently. Airlines are successful with a variation of this by upgrading travellers to gold cards offering lots of perks, then platinum cards offering even more perks, as they travel more frequently during a given period. Make the next purchase increasingly more valuable to the customer. This locks the customer in and increases switching costs over time. It also reduces rewards paid to casual buyers, which then permits even greater rewards to frequent buyers without increasing the overall marketing budget.

Merchants prefer to offer rewards that enhance the value of their product or service. An airline's frequent flier programme enhances the value of the airline because passengers can upgrade for free, or travel with their spouse for the price of a single ticket. Merchants that give paper punch cards already follow this guideline intuitively. A sandwich shop will typically offer a free sandwich after buying several. In some cases, offering a related product like a free drink or dessert could be an even better way to enhance the value of the core product. Programmes offering rewards for other companies' unrelated products or services might sometimes have a promotional impact on customers, but they often have questionable loyalty impact. Some programmes might even damage the merchant's brand.

ENCOURAGE MERCHANTS TO DO THEIR OWN MARKETING, WITHOUT REQUIRING YOUR HELP

Card issuers that create a marketing campaign around concepts like cash back and surprise gifts can focus on communicating the details of their points programme, while letting individual merchants come up with their own ideas for in-store promotions, welcome gifts and other surprises. You don't need to be involved in helping merchants define every coupon or marketing message printed at the bottom of the receipt. If a merchant wants to offer a special discount after a customer's fourth visit, let the merchant decide whether or not the offer will be advertised in the store or simply printed on the receipt, with no other advertising. The card issuer should not feel obligated to provide customers a detailed booklet with all the possible offers available. At most, the card issuer's booklet would include a list of merchants providing special perks and privileges.

This keeps marketing complexity to a minimum. It leverages the marketing creativity of literally thousands of merchant partners, at no cost and effort to the bank. If you own the enhanced payment acceptance brand that is used by the merchant, then you can define rules on how the brand is used and perhaps the types of promotions which are allowed. Some brands might want to disallow promotions on alcohol for example, for a predominantly Muslim clientele, or excessive promotion of bar and restaurant incentives during Lent, in a traditionally Catholic country. If the card issuer has licensed the brand from another party, they will need to be certain from the beginning that the brand fits the issuer's culture and will not clash with the issuer's overall positioning. This should not be difficult, as successful brands will be positioned merely as a payment enhancement feature, with little or no underlying marketing promise other than the commitment that the brand will be widely accepted across many merchant categories in precisely the same way that general-purpose payment brands like Visa and MasterCard promise nothing less than a commitment to provide massive worldwide acceptance. Additional promises on top of that are provided by each bank's unique brand.

If you are an acquirer trying to win business from a leading restaurant chain, you might want to prepare for your next meeting by finding a few coupons that the chain distributes in newspapers or inside the restaurants themselves. Here are three that were in a leaflet the other day, at a restaurant chain in London. One coupon said, 'Buy one main dish, get a free dessert.' The second said, 'Buy one main dish, get a free kid's meal.' The third said, 'Buy one meal, get the second free.' All the coupons had a little additional wording, such as 'only one coupon per visit' and other such details. Try taking those coupons with you to your next meeting with the restaurant chain. Pull them out just as you start describing how your new payment service can help the restaurant chain target their promotions better. Suggest that the restaurant could take these exact same promotions and target each of them based on different criteria. The least expensive coupon, for the free dessert, could be printed at the bottom of the customer's payment card receipt at the customer's first visit. The next coupon, for the free kid's meal, could be printed at the second visit in the same month, while the most expensive coupon for a free meal might be printed only at the customer's third visit during the month. There is nothing complicated here from a marketing perspective. All of the major marketing work is already done by the merchant, in terms of the types of offer, the value of the offers, the wording, the timing and so on. All you are doing is providing the merchant with a payment infrastructure which lets them link their existing offers to each individual customer's payment history.

Enhanced payment capabilities should not require special marketing expertise within the card issuer's organization, nor within the merchant-acquirer's organization, nor for that matter each individual merchant's organization. Each can look at enhanced payment as a way of executing their current marketing strategies better and making their current marketing techniques more effective.

SHOW MERCHANTS HOW YOUR NEW PAYMENT INFRASTRUCTURE HELPS ADDRESS MAJOR PROBLEMS

If you are an acquirer, preparing for an important meeting to pitch your payment services to a major retail chain, you already know the underlying theme that will drive all the negotiations with this retailer. It's the same underlying theme with every retailer. Card fees are too high. And they are growing too fast. You know that even if during the meeting the retailer might be too polite to say so, he nevertheless feels that the payment industry is extracting rent from merchants through merchant services charges that are higher than are warranted by their costs. You prepare to deal with all of these issues. You prepare to explain about the value of lower fraud, efficiencies and new technologies. Perhaps you bring along a representative from Visa or MasterCard who explains the value of brand standards and maybe unveils a national advertising campaign which will certainly help drive business to the merchant. The merchant finds all of this very attractive. Still, he complains that card fees are too high. The retailer explains that card fees are now his fourth-largest expense at the store level. He projects card fees to exceed store utility costs within the next five years and approximate the cost of store rent within ten years. He explains that in fact what he is seeing is that card fees are now approaching his average per-store profits. Any reduction of card fees will automatically go to the retailer's bottom line.

In other words, you are stuck in a cost reduction discussion. You're an expense which must be driven down. Worse, you are boxed in as a provider of commodity services on a par with utilities and rent!

This book has been about breaking out of that box. It has been about getting merchants and cardholders to see payment as something new, exciting and different, so that you can focus on the added value you provide, as opposed to the cost you represent. It has been about transforming your payment infrastructure into a tool that addresses major problems which keep merchants awake at night – problems so big that the costs of your services appear ridiculous in comparison. If you can turn the discussion around and address such issues, you will succeed in radically transforming your business.

The retailer's pain is a rich source of discussion. A little homework before going into a meeting can do wonders. Let's take the example we saw earlier, on convenience stores. We saw how fees charged by card companies on a gallon of gasoline are often greater than the profits earned by the retailer selling the gasoline. We can stay focused on that discussion, or we can move the discussion to other topics. A quick search on the web provides lots of potential issues to discuss:

- Convenience store sales are expected to grow at an average annual rate of 3.8% through 2006, much slower than the rapid pace of 11.3% per year for the past five years. Reports cite heightening competition, declining margins and a heavy reliance on two core low-margin categories (gasoline and tobacco) as the main challenges facing the industry.[53]
- Other types of retailer are increasingly encroaching on the gasoline business and are expected to double their market penetration in the next five years.

- Supermarkets, drug stores, discounters and even fast food restaurants are getting into the convenience game, adopting core convenience characteristics and targeting the same convenience shopper.
- Profits are under pressure as margins get squeezed both at the pump and inside the store. The low-margin cigarette/tobacco category comprises a growing share of in-store sales. Unfortunately, the segment faces declining long-term demand.
- The industry is consolidating around a smaller number of major brand names, and the big players are building brands at multiple levels of the organization, from the gasoline pump to the store to the inside offer.
- Due to all of these elements, slower growth, encroachment by other retail categories, margin erosion and consolidation, convenience stores are concentrating their marketing efforts much more on increasing same store sales than in the past. One survey shows that around half of retailers expect to increase their spending on advertising and promotions in 2005.

What role can your payment services play in these areas? How can acceptance of your enhanced payment brand, and access to the millions of customers that your brand represents, help the retailer address these issues? How much does the retailer currently spend on all marketing activities, including advertising and promotions and things such as free samples and coupons? The retailer could easily be spending 5% of total sales on marketing. Some chains require franchisees to contribute 2% of gross sales to a chain-wide advertising fund and another 1% of gross sales on advertising and promotional activities in their local geographic area. That's 3% on advertising alone, without counting promotions which can quickly add up to the same amount as advertising. A long discussion about how your payment services can help make these activities more efficient would be a rich topic for a meeting.

Discussing major problems faced by merchants will almost always reveal nuggets that you can grab and turn into gold. Near the end of the meeting, perhaps the retailer reacts to something you say about margin erosion. When you ask how the retailer is responding to the growing share of low-margin products in his business, he suddenly launches into an animated explanation of how they are about to offer an upscale coffee programme with a major brand – still hush hush wink wink – which everyone in the organization is very excited about. This was unexpected. You had no idea. Even after speaking to this person on a regular basis for the past six months, trying to win his business away from your competition, you had never heard about this project. You jump at the opportunity and suggest that they should switch to your services before they launch their coffee programme so they can offer a free welcome coffee to customers using your card to fill up for the first time at that chain. Get customers to sit down a few minutes after filling up, try the coffee and maybe buy something else in the store. The customer gets the coupon instantly at the point of sale, when they pay for their gasoline, and, since the coffee station is only a few feet away, it's all so simple. There are no coupons to stuff into envelopes, no coupons to remember to bring into the store, no analysing large databases filled with huge numbers of transactions.

When you come out of a meeting with a major retailer in which you successfully moved the discussion away from your fees and instead focused on the retailer's pain, I can guarantee that you will be feeling fantastic. Even if, after the meeting, the retailer has not yet decided how they will use your payment infrastructure to target their promotions and improve their advertising and marketing activities, more than likely the retailer will have decided to choose you as a strategic, long-term payment partner.

Your cardholders will immediately benefit. Once merchants begin creating excitement around your payment brand in their store, your cardholders will quickly see this. They will be encouraged to use your card more often. And if they don't have one yet, they had certainly better get one fast.

Notes

1. 'Banks face £700 million hit over card fees', *Times* Online, 11 November 2004. According to analysts at Morgan Stanley, the credit card industry as a whole faces £1 billion a year in lost revenue, but 'the biggest banks, who control 70 per cent of Britain's credit card industry, would suffer the largest estimated hit on revenue at £700 million'.
2. 'Precious plastic 2005', PricewaterhouseCoopers, November 2005, available at www.pwc.com.
3. 'MasterCard: EMV non-compliance could be costly', *TheEdgeDaily*, 9 July 2004.
4. 'From strips to chips for security', *Australian Financial Review*, 5 September 2002.
5. 'An open invitation to card fraud?', *American Banker*, 1 July 2004
6. 'Card fraud a headache for Malaysia's banks', *Forbes Digital Tool*, 24 April 2003.
7. 'Bankers Still Lying Awake at Night', *Singapore Business Times*, 19 May 2004
8. Testimony of Michael Turner, The Information Policy Institute before the Subcommittee on Financial Institutions, Committee on Financial Services, US House of Representatives, 8 May 2003.
9. *The Nation*, 29 May 2004.
10. 'Credit Card Fees a Growing Challenge for Convenience Stores (Fact Sheet)', National Association of Convenience Stores, www.nacsonline.com.
11. '$3B Credit card suit good for shoppers', CBS News, 1 May 2003
12. 'Competition in banking has reduced business costs – Merchants paying less in merchant service fees on credit cards', *Australian Bankers' Association*, 15 July 2004.
13. 'Rewarding new card from NAB and Amex', *The West Australian*, 14 July 2004.
14. 'Banks face £700 million hit over card fees', *Times* Online, 11 November 2004
15. 'Card growth in China drives merchant fees down', *China Daily*, 2 February 2004.
16. 'The merchant acquiring business: Commodity service or opportunity for business growth?', Edgar, Dunn & Company, July 2004, *Insight*, vol. 11, also available at www.edgardunn.com.
17. 'First Data bets on cards for epayments', *Cards International*, 20 January 2003.
18. 'Nova ramps up in Europe with 3 deals in a week', *American Banker*, 3 May 2004
19. 'Profiting from EMV: Bad debt provision and new revenue streams', Jane Adams on behalf of ACI Worldwide.
20. Press release, 'Mashreqbank Q3 2004 profits climb 22.3%', 9 October 2004
21. 'Niche marketing of credit cards pays off', AME Info, United Arab Emirates, 2 August 2004.
22. 'The customer experience', *Fast Company*, Fall 1999, 1.
23. 'White Paper – Engineering customer experiences', IBM Advanced Business Institute, 12 October 1999.
24. McKenna, Regis, *Real-time: Preparing for the age of the never satisfied customer*, Harvard Business School Press, Boston, MA, 1997, p. 56.
25. 'Your mileage may vary: Why frequent fliers are often grounded', Michael Shapiro, *The Dallas Morning News*, 22 August 1999, p. 1G.
26. 'Boots launches loyalty card scheme', M2 PressWIRE, 11 August 1997. Also: 'Boots shows its hand and joins card wars', *Independent*, 7 August 1997.
27. *New Rules for the New Economy*, Kevin Kelly, Viking, Harmondsworth, 1998, p. 14.
28. *Complexity: The Emerging Science at the Edge of Order and Chaos*, M. Waldrop, Simon and Schuster, New York, 1992.
29. 'Institutions in the Age of Mindcrafting', Hock, Dee W. (1994), speech presented at the Bionomics Annual Conference, San Francisco, California.
30. Cards and Loyalty Conference, Cartes 2004, Paris, 3 November 2004.
31. 'The competitive dynamics of network-based businesses', Coyne, Kevin P. and Dye, Renee, McKinsey & Co, *Harvard Business Review*, 1 January 1998, p. 99.

32 'The one-horned cow', Hock, Dee W. (1994), speech given to the Graduate School of Bankcard Management, Norman, Oklahoma.
33 *Out of Control: The New Biology of Machines, Social Systems and the Economic World*, Perseus Books Group, Reading, MA, 1995.
34 *The History of Credit Cards*, Gareth Marples, Net Guides Publishing Inc., Las Vegas, 2004.
35 'Tricks of the POS Trade', *Card Technology*, November 2004.
36 'Multi-application terminals in a changing payment environment', Multi-Application White Paper, Verifone, June 2002.
37 'Smart cards, payment terminals, and the point-of-transaction', *Hyperline*, a publication of Hypercom Corporation, Issue 1, 2003.
38 'Message from the Chairman', Ingenico's 2002 annual report.
39 STIP Consortium website, www.stip.org, About STIP, November 2004.
40 'Dee Hock on management', *Fast Company*, Waldorp, M., October–November 1996, 5, 79.
41 *Wind, Sand and Stars*, 2nd edn, A. de Saint Exupery, Harcourt, Orlando, FL, 1992.
42 *Built to Last: Successful habits of visionary companies*, HarperBusiness, Collins, James C. and Porras, Jerry L., (1994), pp. 142–3.
43 In the Future ..., MasterCard International corporate brochure, September 1997.
44 Relationship Card, Visa Corporate Relations, corporate brochure, 1995.
45 'Sainsbury to cut headquarters jobs, halve dividend', Bloomberg news release, 19 October 2004
46 'Revealed: the secret of Tesco's success', *Times Online*, 18 October 2004.
47 'Private label initiatives drive new competition in the store card market', Mercator Advisory Group, 14 October 2004
48 'The co-branding boom', *European Card Review*, November–December 2002.
49 Proton Statistics: www.proton.be, and 'VISA: Major smart card players create proton world international', M2 PressWIRE news service, 30 July 1998. Carte Bancaire statistics (1997) '5 années de cartes bancaires à puce', *Guide de la Carte, Analyses & Synthèses*.
50 'Smart cards need to be smarter', Veverka, Amber, Knight Ridder News Service, *San Jose Mercury News*, 3 September 1999
51 'Jupiter Research reports US online retail sales will reach $65 billion in 2004', press release, 20 January 2004.
52 Smart card talk newsletter, Executive Director's Letter, Smart Card Alliance, March 2004.
53 'Heightened convenience competition hinders projected c-store sales growth', *Retail Forward*, 16 January 2003.

Index